ANCIENT WOMEN GARDENERS

New Century Gardens and Landscapes of the American Southwest
BAKER H. MORROW, SERIES EDITOR

Whether practical gardening guides, best plant guides, landscape architecture showcases, or blueprints for urban ecology, books in the New Century Gardens and Landscapes of the American Southwest series address the challenges novice gardeners and skilled practitioners alike face with prolonged droughts, limited water supplies, high-altitude climes, and growing urbanization. Books in this series not only provide practical landscaping advice for backyard gardeners, they dive deep into ecology, built environments, agricultural history, and the emerging discipline of urban ecology. The New Century Gardens and Landscapes of the American Southwest series tackles the environmental questions that many communities in the American West confront as we all work to create healthy, dynamic, and inviting outdoor spaces.

Also available in the New Century Gardens and Landscapes of the American Southwest:

The Design Competition in Landscape Architecture: Pedagogy and Practice
    by Kathleen Kambic and Katya Crawford
Growing a Sensational Garden in the Southern Rocky Mountains: A Monthly Guide
    by Nan Fischer
Feeding a Divided America: Reflections of a Western Rancher in the Era of Climate Change
    by Gilles Stockton
The Gardens of Los Poblanos by Judith Phillips
Water for the People: The Acequia Heritage of New Mexico in a Global Context
    edited by Enrique R. Lamadrid and José A. Rivera

ANCIENT WOMEN GARDENERS

Prelude to the Chacoan World

DAVID E. STUART

UNIVERSITY OF NEW MEXICO PRESS | ALBUQUERQUE

© 2025 by David E. Stuart
All rights reserved. Published 2025
Printed in the United States of America

ISBN 978-0-8263-6846-1 (cloth)
ISBN 978-0-8263-6847-8 (paper)
ISBN 978-0-8263-6831-7 (ePub)

Library of Congress Control Number: 2025006773

Founded in 1889, the University of New Mexico sits on the traditional homelands of the Pueblo of Sandia. The original peoples of New Mexico—Pueblo, Navajo, and Apache—since time immemorial have deep connections to the land and have made significant contributions to the broader community statewide. We honor the land itself and those who remain stewards of this land throughout the generations and also acknowledge our committed relationship to Indigenous peoples. We gratefully recognize our history.

Cover illustration: courtesy of Baker Morrow
Designed by Felicia Cedillos
Composed in Adobe Caslon Pro

I dedicate this book to Anacelie Verde-Claro, poet, long-time friend, and transcriber of my handwritten manuscripts.

Contents

Acknowledgments ix

Introduction 1

PART I. HUNTERS & FORAGERS

Chapter 1. After the Ice Age 15

Chapter 2. Changes in the Four Corners Region During the Early Archaic Period 35

Chapter 3. The Robust Plant and Foraging Society of the Middle Archaic Period 53

PART II. CORN & WOMEN'S GARDENS

Chapter 4. The Genius and Innovation of Ancient Women's Pocket Gardens 69

Chapter 5. Farming Labor Patterns 91

Chapter 6. The Evolutionary Balance Scales Shift, 1–600 AD 99

Chapter 7. From Basketmaker to Puebloan 117

PART III. SMALL FARMERS & GREAT HOUSE ELITES

Chapter 8. Geography and Chacoan Society's Emergence 131

Chapter 9. The Dynamics of a Semiarid Empire 149

Chapter 10. The Great House Era, 875–1175 AD 157

PART IV. CHACO CANYON'S DOMINION & FALL

Chapter 11. Modifying Landscapes as Avenues to Power   177

Chapter 12. The Rhythms of Great House Power, 900s–1100s AD   189

Chapter 13. Tiptoeing on the Edge of Chaos, 1130 AD   201

Epilogue. Pueblo Women: Past, Present, and Future in Knowledge, in Change, in Balance   217
   ELIZABETH AKIYA CHESTNUT

Postscript   223

Notes   225

Bibliography   233

Index   239

Acknowledgments

I owe a debt of gratitude to Bill Dunmire for his advice on native plants, Hayward Franklin for ceramics information, and Elizabeth Chestnut for her conversations with me about the roles of Pueblo women. My thanks also go to Esther Burton, Dawn Davis, Jesse Cochran, cartographer Morgan Hite, University of New Mexico Press Senior Editor Sonia Dickey, series editor, professor, and noted landscape architect Baker Morrow, supportive friends Ian Antonio Ellis and "Hank Tee," and noted editor Brian Weiss.

I am grateful to my Stuart great-aunts who taught me the thirty-six generations of our clan lineage when I was a boy living in my grandfather's multigenerational household on Bainbridge Street in Philadelphia. They taught me that the Isle of Bute in Scotland was my true "home," as well as our complicated family history dating back to the mid-900s AD in Brittany. I inherited a living lineage, its behavioral lessons, and detailed cultural histories. I did not just inherit a surname—I was part of a living human chain tied to a cultural and historical reality.

I am indebted to the many field archaeologists who, in the late 1970s, refined our knowledge of archaeological elements tied to the Chaco Phenomenon. Among them are Richard W. Loose, Michael P. Marshall, John R. Stein, Rory P. Gauthier, Thomas Windes, Nancy Akins, Joan Mathien, and the brilliant Cynthia Irwin Williams, whom I first met about 1965 as a guest lecturer in Mexico City during my student days at the University of the Americas.

This book was written in its entirety at warm, friendly Limonata Café on the corner of Wellesley and Silver in Albuquerque, New Mexico. Writing there reminds me of the great writers I used to meet and chat with at the Café Punto Blanco in Mexico City. Among them was the famed Dr. Octavio Paz, author of *Labyrinth of Solitude*, for which he won a Nobel Prize. Thank you for your example and your generous encouragement, don Octavio.

INTRODUCTION

THIS BOOK BEGINS WITH the climatic, ecological, and cultural changes in the North American Southwest that eventually led to the rise of a complex regional society centered on Chaco Canyon, its Great Houses, engineered roads, marvelous wash district, and thousands of widespread gardening hamlets.

We first discuss ancient hunters-foragers, then move on to the long era of broad-spectrum foragers, then meet the first women gardeners who set the stage for a later complexly tiered world of Chaco Canyon farmers, potters, master masons, artisans, powerful women landowners, a few hunters, and an elite, ruling priesthood. The cultural life spans of each distinct form of these successive societies were larger, more complex, and shorter-lived than their immediate predecessor society. This has much to do with energy flows, biological reality, and the mathematics of "power" and "efficiency." It also sheds light on structural and energetic dynamics that archaeological texts rarely address.

Yes, calories, a measure of heat and energy, played an essential role in the nature and succession of Southwestern societies. According to Alfred J. Lotka's mathematical laws of human population growth and the famed "Darwin-Lotka Energy Law," there is an evolutionary advantage to organisms—or cultures, in this case—that can maximize energy flow (Lotka 1922). According to Lotka's ecological thermodynamics, when a society needs to transform and restore a given amount of energy at the fastest possible rate, some 50 percent of it must be lost due to the work invested. That loss is precisely what happened in Chaco Canyon about 870 AD, a dynamic that shaped the energetics of the next three centuries of a rapidly and increasingly complex Chacoan world that eventually crumbled, as it never overcame its early energy deficits from the investment in dozens of huge masonry Great Houses.

Chaco society's origins lie in the "Greater Southwest," which includes parts of Arizona, New Mexico, Colorado, Utah, and the Mexican states of Sonora and Chihuahua. Small, family-based bands of ancient hunters and foragers roamed all these lands as of about ten thousand years ago. Among these, some drifted north from Mexico, possibly pushed by the combination of population increase to the south in west-central Mexico and an increasingly dry and barren district in what is now northern Sonora and western Chihuahua. Parts of what are now eastern California, Arizona, and western New Mexico offered a somewhat moister climate, more extensive grasslands (harvested by hand for seeds) and protein-rich plants like Indian ricegrass, wild amaranth, and spinach-like *chenopodia* species. These plants provided essential dietary protein and carbohydrates. In ecological terms, both water sources and plant diversity were rapidly declining in western Chihuahua and northern Sonora during this period. It was time for humans to move on to denser vegetation and cooler temperatures to the north. This ancient foraging life way in the Southwest dominated until about thirty-five hundred years ago, when domesticated garden crops like small-cobbed corn, beans, and squashes became available to descendants of those Sonoran and Chihuahuan migrants and to even older resident native populations of the Southwest (MacWilliams 2018).

The greater Southwest had supported modest populations of hunting-foraging peoples, originally of northeastern Asian descent, since the end of the Ice Age. Alaskan and western Canadian populations are well-dated from child burials to at least eleven thousand years ago (Raff 2022), and possibly as early as sixteen thousand years ago. Among their Indigenous Alaskan and Canadian descendants were the forebears of the ancient Native peoples who still live in the modern Southwest.

Genetic DNA and human language patterns have long lives that allow analysts to trace ancient origins and relationships. Based on female mtDNA which unlike male Y DNA does not divide, there is no doubt of direct descent of ancient populations in Alaska, Canada, and the northwestern United States, where Native peoples have occupied the same regions for nine to eleven thousand years, and longer. Though some ancient peoples migrated to less occupied areas as populations increased, oral history helps to remember original American Indian land ownership, which is certified by DNA.[1]

Nine to eleven thousand years seems an unfathomable eternity to decipher daily behavior, marriage patterns, religion, rituals, languages, and group identities. Field archaeologists rely on findings such as fire-fractured rocks, stone flakes, a fragment of animal bone, or the charred grass seeds and pollen traces of the Southwest's Early Archaic Period circa 6000 BC, all carefully sifted from the hearth soil of an ancient windblown sand dune camp in west-central New Mexico (Vierra 1994).

Most scholars of the Southwest would assert that those nine thousand years before the "Common Era" began with large game hunting and subsidiary foraging behaviors of widely scattered late "Paleo-Indian" peoples, followed by "Archaic" peoples who hunted smaller game and foraged plants. Following this was an era of innovative female "Basketmaker" gardeners, then early Puebloan farmers. The Chacoan version of Pueblo society peaked in the complex era of the 1100s AD with the fall of fabled Chaco Canyon Great House society and its scattered Puebloan survivors. But the Southwest's slow-motion saga of those nine thousand years is far more subtle and complex than any relatively modern historical narrative—even that portrayed in Edward Gibbon's famous multivolume work, *History of the Decline and Fall of the Roman Empire* (Gibbon 1996).

Written in the late 1700s, Gibbon's epic work is revered for its long view of Rome: its rise to power, its success, its episodic follies, followed by multiple reorganizations, the hubris of its own power, and ultimately its demise at the hands of an increasingly rich, crazed, incompetent, complacent, isolated, and insatiable ruling class—most of which was repeated in smaller scale by Chacoan Great House society a continent away and a thousand years later.

Scholars and the public alike obsess over Chaco Canyon, its Great Houses, and Mesa Verde's stunning cliff dwellings. Both still radiate the silent but seductive—and culturally familiar—glow of the power they once generated. Yet those two archaeological societies' long-gone peaks of power had agency, or social power, for only five to six centuries... at most a percent or two of the story that led from ancient foragers to the confederacy of efficient female gardeners that enabled the formation of Chacoan, Mesa Verdean, and other complex regional Native American societies.

Efficiency

To fully understand most of the Greater Southwest's historical dynamic, one needs also to understand the concept—and the protective value—of efficiency. Efficiency factors generate inexorable changes in the human saga and its complex adaptive systems that we call "cultures." Waste too much food or energy too quickly and a period of want ensues. Hunger and want generate modified cultural behaviors.

In the late 1940s a federally mandated efficiency phase was ordered in the American homeland to support the military's efforts in World War II. That deeply shaped my own childhood, just as it shaped the lives of legions of other children born in the 1940s. I remember automobiles sitting up on blocks in nearby backyards. Tires were unobtainable. I remember the garden work, the organized female "cannings" at harvest time, the flour-sack school clothes sewn by my mom and my aunts, and the eagerly awaited Sunday dinners served with home-canned vegetables and our own hens' eggs. Our garden in the 1940s and '50s in the town of West Chester, Pennsylvania, was just about an eighth of an acre, bounded on two sides by huge blackberry bushes. That garden was about the size of the average ancient Anasazi woman's highly diverse "pocket garden," as first planted in the 1100s to 800s BC.[2] Our one garden fed us through the winter. The only purchases we made were cooking oil, flour, and milk.

Make no mistake—a society that remains inert in the face of unexpected change and stubbornly doubles down on "tradition" opens a clear path to eventual cultural extinction. Dramatic increases in complexity in order to solve urgent short-term problems are another potential path to eventual cultural failure, fragmentation, and collapse. Complexity is often an enemy of efficiency, and efficiency is *the* key to long-term cultural survival.

Many of the Ancient Southwest's cultural, technological, and demographic transformations were slow, subtle, and modest enough to go largely unnoticed by early archaeologists. Imagine a small society transformed by the adoption of a slightly better-shaped hand grinding stone, or by an unknown woman who once took wild grass stems and braided them to make a coil for her child to play with. That grass coil may have become the basis of all Late Archaic Period coiled, grass-seed storage baskets. A wide variety of wild grass seeds had been basic food sources for much of the postglacial continental Americas.[3] Grass seeds were high in protein, fiber, and essential minerals. In addition, grasses provided fiber and stems for making cords, rope, and sandals. Much later, corn would evolve from two crossbred wild grass strains.

According to William W. Dunmire and Gail D. Tierney in *Wild Plants and Native Peoples of the Four Corners* (1997), the most nutritious grasses were *oryzopsis hymenoids*, which flourished in many elevational zones and produced 120 calories per ounce of protein-rich seeds. An extended family of ten could subsist on grass seeds for about forty days in early spring. If a few rabbits were captured, that family could subsist for about forty-two days. To restore body weights to the fall norm, 1.5–2 acres of claimed Indian rice grassland was adequate (Dunmire and Tierney 1997).

The storage of excess harvested wild grass seeds changed the Archaic Period's cultural and energetic dynamics and, thus, changed the speed, scale, risk, and trajectory of an entire ancient way of life. Storing grass seeds was efficient, but it also created a potent survival strategy in the face of scarcity during long, colder-than-average ancient winters. Stored food smoothed out the food production of nature's seasons, rather like a stable monthly salary sustains a modern household. Without the eventual widespread storage of edible grass seeds and small-cobbed maize, there never would have been a Chacoan world about a thousand years later.

Even Rome's fortunes rose and fell on the stability and predictability of grain available for its ordinary citizens to make their daily bread. Roman emperors rightfully feared grain shortages at home far more than facing competing armies abroad. That fear is what nudged the Romans to deploy the empire's legions to North Africa to produce wheat and to build its signature power architecture of amphitheaters and gladiator events.

In stark contrast, during the Early Middle Archaic Period in the Southwest, the fear of want fell most heavily on forager mothers. Both subtle and not so subtle natural changes in the landscape, its ecology, and notable climactic vagaries frequently demanded emergency responses and shaped daily food gathering behaviors. An unusually cold winter or protracted scorching summer winds demanded behavioral adjustments, rather like adding the weight of a second penny to one pan of a balance scale: a small weight creates momentum as the lighter pan rockets upward. The weight of the rising pan and the rate of its upward distance produced measurable power. The Early Archaic Period's changing climates and ecological environments nudged fragile, scattered populations to experiment with food gathering and storage adaptations when climatological changes took place.

Over the next five thousand years (roughly 8000 BC to 3000 BC), temperatures rose, ancient ponds shrunk, and shrubby grasslands began to replace once-rich piñón, juniper, and upland oak belts, which reduced the protein-laden bounty of grass seeds, piñón nuts, and acorns. Over-foraged grasslands and the common practice of disturbing grassland soils by repetitive digging for root foods like wild potatoes eventually impoverished soils, rendering them more vulnerable to invasive weeds. Many large game animals, like giant elk and saber-toothed predators, also became extinct as climate heated up and smaller animal species consumed an increasing portion of plant foods. For most human families, the ratio of wild seeds eaten rose notably and came to be a daily staple. Ancient peoples also adapted by moving more often and intensifying their seed harvesting labors. Each seed-based food calorie earned came at a higher labor cost than had high calorie and protein-dense meat. Piñón nuts offered high protein and relatively labor efficient food calories, but piñón trees typically produced crops only every few years. Foraging for them was an on-again, off-again food strategy. In contrast, wild grasses that annually produced seeds eventually became the most important ancient food staples.

Power and Efficiency

The ratio of power to efficiency is easy to observe. Put two equal weights on a double balance scale and the result is inertia—perfect efficiency. Nothing moves. Perfect efficiency generates no energy to actually move the scale. Then add a penny to one pan and watch the lighter scale pan rise. That rise is the visible evidence of work. The work done is measured by the weight of the rising pan as a ratio of the weight of the falling pan and its length of travel, which is heavier because of the added penny. The important concepts of power and efficiency suffuse this book because efficiency-focused societies typically outlive power-focused ones.

In our modern American nation's most deeply confusing times and amid widespread ecological malfeasance, it makes sense to stand back and look more closely at the broad nature of human societies. The perspectives provided by anthropology can be of some assistance. In our species's early societies, small family-based bands roamed their landscapes—foraging, hunting, and scavenging. Each band averaged thirty to fifty individuals and, based on ethnographic research (Stuart 1972), their entire society was numbered in the hundreds—three to five hundred souls was a common average for a given hunting-foraging band and its language. Personal goods were few, and each day's harvest of wild vegetables, birds' eggs, or small game was typically consumed in its entirety and stored largely in the band's collective body mass. Few possessions were owned. Life was, in fact, one long ceaseless round of the same: such societies generated few possessions, the rate of innovation was glacial by modern standards, and population was controlled by ritual, taboo, sexual proscriptions, and the availability of food resources in the band's foraging region.

A number of the longer-lived of these societies survived largely unchanged for five to eight thousand years; but among the cultural champions were a number of aboriginal Australian language communities that had survived for more than thirty thousand years until Captain Cook and the British Empire ruined the delicate balance between forageable land and survival.[4] The coastal Seri peoples of Sonora, Mexico, and the Selk'nam and Yahgan Indians of Tierra del Fuego island at the southernmost tip of South America also deserve honorable mentions.[5] In the last six hundred years, many of these long, stable societies have disappeared. Modernized societies invaded, then complexified, radically altering those ancient territories' long-stable ecosystems.

Far from the "primitive" label given them by their colonizers, these ancient societies represented paragons of efficiency: they consumed little, wasted little, and typically maintained their populations at sustainable numbers. The rights to marry played a powerful role in population control, as did cultural practices surrounding menstruation and childbirth. These traditional societies were also technologically stable, innovating when population, resources, and changing weather rendered things out of balance—absent the unexpected, which included intrusions from modern societies, the least complex and most efficient verged on immortal. There are no large, complex societies that have matched these ancient hunter-foragers' cultural and demographic staying power.

Of Dynamics: A Modern Example

The contemporary United States represents quite the opposite of stable cultural efficiency. The United States has become a power- and change-obsessed raging bull of profligate consumption, stunning waste, and extraordinary complexity. According to European researchers, the United States is now even *more* economically unequal than was Nero's Rome. Rome did not fall because of its moral "decadence"—it fell because its growing complexity and quest for power over an expanding geographic sphere was unsustainable given available resources.

In the real world, every society metabolizes as it fluxes resources and expands or contracts, just as the Chacoan world once did. In fact, I argue that human cultures have been influenced by our own daily body rhythms and the cycles of night and day. Each of us metabolically repeats the same processes every twenty-four-hour cycle: we cool down, quit consuming calories, and go into sleep mode at night. That is when body temperature drops, brain activity outpaces muscular activity, and body repair programs go to work . . . this exemplifies a built-in corrective "efficiency phase" in each of our daily cycles. Biological efficiency is, literally, in our genes. In contrast, economic efficiency is a product of culture.

Real History

In fact, as I argued in my last book (Stuart 2019), the American nation and its public culture was actually founded on the biological and economic advantages that its European colonists enjoyed from about 1650 to 1880 at the expense of Native American populations.

However, American longevity has been declining ever since the late 1970s; the advantageous biological and economic parameters that led to the Euro-American nation's birth and stunning growth are currently fading. The American nation is now measurably more unequal in economic and human body conditions, and *far* less efficient than it was in the 1950s and '60s.

Every society goes through the struggles of metabolic-like phases of cycling ups and downs. Societies typically measure these trends through surrogates like births, deaths, wealth, and want. What does this all mean as we ponder the people of the Ancient Southwest? It requires us to think about the daily lives of the region's bands of hunters and foragers, to focus on the subtle details of their ecology, technology, and innovations. It requires us to respect subtlety and whispers of change . . . in an era when there were no breathlessly manufactured crescendos like our daily "breaking news" flashes. On most ancient days, drama was the sound of a rabbit crossing a nearby sand dune, a sudden cool breeze, a turkey buzzard slowly circling above, or a migratory butterfly appearing out of season.[6]

Cycles of Power and Efficiency

The long-term rhythms of human societies are complex, especially in large, modern societies.[7] In the course of an ordinary human lifetime, currently seventy-six to seventy-seven years in the United States, our society will have gone through several distinct phases: spurts of rapid growth, then slowdowns or outright retrenchments. We tend to think of these cycles in the associated surrogate terms common to the fields of economics, technological history, demography, and politics. As a nation, we tend to blend all these together into the alchemic folklore that we view as "history."

To me, an ecological and cultural evolutionary anthropologist, the oddity of this is that our standard histories are merely a shorthand mélange of the rhythmic statistical symptoms of interactions between variable energetic states (energy, flow, and distribution), demography (birth rates, infant mortality, height, longevity, etc.), and the political, economic, technological, and social responses to these systemic changes.[8] Yes, average height, longevity, and infant mortality in a society's population are important keys to understanding both the distributions of food/wealth and "agency" (cultural power) among a society's citizens. The statistical methods that evolutionary economists rely upon are both crucial and hard to understand. One of the best and more understandable of those scholars is the late Robert W. Fogel, a Nobel Prize winner and former professor at the University of Chicago. One of his books, *Explaining Long-Term Trends in Health and Longevity* is useful in explaining our biological rhythms (Fogel 2012).

How are Contemporary Americans Doing?

The size of a culture is also important. The United States is so large, so regional, and so class/race-based that not every community experiences the overall state of the nation—instead, a number of distinct American subpopulations (Black Americans, immigrants, Native Americans, Hispanics, and the poorly educated) are too often politically or corporately manipulated and thus do not have a clear picture of either their role in society or how the nation works. This is why education, work experience, and skills really do soften the negative effects of "not belonging."

"How are we doing?," a simple question in the modern United States, therefore becomes a staggeringly complex puzzle in a complex and fragmented world. In later chapters, we will see these patterns emerge in the ancient San Juan Basin during the run-up to the Chacoan world as it grew in pulses between 750 and 1130 AD, complexified, stratified, ossified, and then fragmented. That complex Chacoan world had only emerged *after* more than two thousand years of increasingly sophisticated gardening (typically a female role) and corn horticulture (typically a male role) extensive enough to classify as "farming."[9] In the process, an elite class of priests and residents gained disproportionate power and hid their massive stored food supplies from the general public in cloistered Great Houses.

At the height of most cultural power phases, crucial cultural and economic information is sequestered. Only the priests, CEOs, or intelligence agencies know the facts of each other's specialized power niche and their institutions. Thus, resources are often foolishly distributed simply to maintain the fiction of near-absolute power and control. Wanton waste and treachery are applauded as "rights" held by a few. One can label the society as "empire," "regency," a "power phase," or the "Chaco phenomenon." Chacoan society, at its height, was *not* egalitarian. Indeed, its "underclasses" were typically shorter, leaner, and died younger than the Great House elites and those in the priesthoods. History tells us that it is neither the most powerful nor the most profligate societies that will inherit the earth . . . it is the most *efficient*.

Many of the powerful societies known to historians, ethnographers, and epigraphers have been replaced over time by more efficient successors able to profit from the abandoned infrastructure and vacated lands of a former empire. That, too, is part of the Chacoan story and the later saga of Puebloan and Diné (Navajo) peoples who geographically replaced many once-vibrant yet scattered Pueblo communities created in the thirteenth and fourteenth centuries AD. Now, we must turn back to the cultures that preceded the Chacoan world.

PART I
HUNTERS & FORAGERS

CHAPTER 1

AFTER THE ICE AGE

ARCHAEOLOGICAL FIELD RESEARCH AND refined dating techniques conducted in the last several decades have pushed back the possible dates for the first peopling of the Americas to roughly sixteen thousand years ago (Raff 2022). Early peoples from Eastern Asia and Siberia arrived in an ice-free part of Alaska during a period of substantial glaciation. To the south and east of these first peoples, slowly melting snow fields, extensive cool wetlands, countless lakes, and a suite of very large Pleistocene animals awaited their descendants: wooly mammoths, horses, immense cave bears, giant sloths, and saber tooth tigers. Meat, if they could hunt or scavenge it from other predators, was plentiful. So were fish, like salmon.

Those who first arrived in Alaska from Asia ran into barriers of huge ice fields and deep snows in what is now Alaska and western Canada. For a time, terrestrial groups of humans were trapped by glaciation in interior parts of Alaska, but others who came in skin boats or canoes stuck to the Pacific coast and moved far more freely than those in the colder, glaciated interior (Dixon 1999).

In 2011, one amazing archaeological project uncovered the cremated remains of several children in an ancient camp in the Upward Sun River district of Alaska. One of the female infants became known as "Dawn Girl." Her female mtDNA allows us to know her as a direct ancestor of living Native American–language groups still inhabiting the region where she was buried. "Dawn Girl" matters to me. Her careful burial shouts love, empathy, and sorrow across the ages and also proves an irrefutable female genetic thread confirming that the current Native American populations in the region of her burial are her mother's direct female descendants, still subsisting on lands they have held or used for about eleven thousand years. Dawn Girl's people were efficient users of the land, as are her descendants (Potter et al. 2011).

Among the distinct coastal dwellers with Asian origins who came to the new world were peoples whose ancient DNA groups have left definite genetic tracks in modern DNA samples taken from a variety of Native American populations. Those populations range from the North American Arctic to the uplands of Chile (Dillehay 1997) and to the cold, windy, southern tip of Tierra del Fuego Island in South America. Humans have been on the New World's species list far longer than scholars imagined fifty years ago.

In their seagoing boats, these early Asian transplants settled and "mapped" themselves onto the diverse landscapes they encountered on their way south, with the sea on their right shoulders and the shore on their left. Both marine and terrestrial animal life was abundant. In places where fresh water flowed into the seas, some seafarers made coastal homes. Eventually, however, both human impacts and a warming world climate reduced the number of the Ice Age's giant animal species. Immense lakes shrank as temperatures warmed and became seasonal ponds called *playas*.[1] Over time, the warming climate trend eliminated many of the mastodons, huge cave bears, and armadillo-like ground sloths the size of a hippopotamus. Those species were replaced by horses, camels, and herds of huge bison (*bison antiquus*) that flourished in rich expanses of wild grasses and that lingered a bit longer than the true mammoth giants and early apex predators (such as saber tooth tigers). Some scholars have argued that human predation diminished these species lists (Martin and Wright 1967), but as scientists now understand it, there were simply too few hunters to dramatically modify North America's late Ice Age biosphere.

The upside of postglacial warming included the expansion of trees, plants, and grasses. In much of the greater Southwest, piñón, juniper, and oak trees all provided easily stored, high-protein seeds and nuts. By 9000 BC meat was also plentiful, and human populations small.

Rising solar heat, along with the time needed for plant and animal species to adapt, generated a natural power phase in the biotic world. Then gentler and warmer postglacial climate patterns encouraged human population growth. As the earth warmed and the cold seasons shortened, less human caloric/metabolic energy was required to maintain healthy body weight and temperature. Other caloric savings came from the declines in work energy needed for fuel gathering and fire-tending. The amount of time spent on tedious hide processing for heavy winter pelt parkas also declined. Reduced parka weight also reduced calories expended on short treks. Even the high caloric cost of walking in deep snow and on slippery ice declined dramatically. Finally, the intimate task of keeping infant children warm was reduced. Fewer of mom's body heat calories were required. All these saved calories derived from a major environmental shift to warming air temperatures and favored modest population growth.

As the world of snow and glaciers waned, the heat added by postglacial climatic changes did much useful "environmental work," which benefited the small, scattered groups of humans who were quickest to capitalize on more abundant plant and small animal food species. As moderating climates became more seasonal, those family groups also became habituated to roaming over more land in the course of each year's weather cycles. Thus, emerging human work, travel, and food gathering became more seasonal.

In the greater Southwest, the ancient worlds of the Ice Age gave way to warmer climates, smaller animals, more edible plants to exploit, a few more human births, and an era of adapting to a rhythm of life composed of a series of increasingly shorter seasons. Each lived season encouraged the ancient human groups to both target different resources and refine the specialized technology of tools, cooking procedures, and food storage. Each season's distinctive plant and animal food resources created a relatively stable rhythm. It became a warmer, gentler world in which cold and ice no longer so dramatically restricted human movements and activities. This era supported Paleo-Indian groups like the southern plains' Clovis buffalo hunters of New Mexico and west Texas in the period of about 9500–8500 BC (Vierra 2018).[2]

These postglacial cultural adaptations are generally referred to as Paleo-Indian (Ancient Indian). The Clovis Paleo-Indian way of life focused on big-game hunting of species like elk, bison, mule deer, and desert goats. Bison was the favorite; multiple families often cooperated to kill and butcher as many as ten large bison. That bounty supported larger camps and led to regional rises in population; it also required annual access to tens of thousands of acres in which to systematically hunt and gather. Most sources label these meat hunters and seasonal foragers as a version of an ancient band society. A kin-based band society that spoke their own language likely included one hundred to four hundred people. Each band claimed rights to forage and hunt and typically defended their hunting-foraging domain's boundaries from interlopers. This lifeway endured for thousands of years in the greater Southwest. Its accomplishment resulted in exquisitely made *atl-atls*, finely flaked dart points, well-made bison hide scrapers, and woodworking tools.[3] Meat and animal organs and fat were the preferred foods, but the women often foraged for plants.

Much research into Paleo-Indian adaptations remains to be done. Modern laboratory analyses, increasingly sophisticated dating technologies, bone-derived nitrogen ratios, and advances in genetics are yielding up more information on ancient diets, hunting practices, and foraging activities. Analysis of ancient coprolites (desiccated human fecal remains) can yield clues to diet, MT (female) and Y (male) DNA and, with luck, blood groups. Analysis of human bones and teeth can yield gender, age, the region where an individual was raised, type of food eaten, nutritional disasters in infancy, and medical conditions.

A very ancient example is provided by the analyzed remains and preserved coprolites of Paleo-Indian inhabitants from the Paisley Cave complex in Oregon. These caves were inhabited by about 8500 to 9000 BC. The ancient cave's residents ate a variety of foods, both plant and animal. Animal meat dominated, but wolves, coyotes, wolverines, and other predators were not eaten. Surprisingly, several swirls of loosely twisted bulrush stalks were found even though the Paisley Cave complex was located in a very dry area. This suggests that the bulrushes were gathered and brought from a water source or soggy meadow at some distance from the caves.

Though no human remains were recovered (save for the coprolites), stone tools were found. The nature of the Paisley lance heads and their fragments was technologically distinct from those of

Bulrush sketch by Esther Burton.

the American Southwest's Clovis assemblages. For many years, Southwestern archaeologists assumed that the Clovis people were the *only* ancient lineage of North American Paleo-Indian peoples, but the Paisley Cave lance heads were much thicker and were flaked by a quite different method than Clovis tools. The carbon and isotope dated materials in the Paisley Cave complex also indicate that this cave and its technological traditions *pre-dated* the earliest Clovis sites by about a thousand years. Now known as the Western Stemmed Point Tradition, it was made by a society at least as old as Clovis (Raff 2022).

Of course, the Clovis tradition's most ardent archaeological advocates have been skeptical of earlier archaeological sites. But archaeological finds from Alaska to Tierra del Fuego all point to multiple ancestral technologies, lineages, and adaptations to varying climate and landscapes. We humans in the Americas are also more diverse than many scholars once assumed.

Here in the Southwest, successive, technologically distinct Paleo-Indian cultures existed from about 11000 BC to about 6000 BC. Waves of major climatological changes (cold and icy, warm and dry, then much colder) were followed by ever more distinct seasonality and less predictable precipitation patterns. Of necessity, ancient populations slowly but inexorably transformed as some ancient Southwestern Paleo-Indian societies increased their plant consumption and began to forage more widely. Based on isolated dugout food caches that have been found, we know that ancient foragers began to engage in modest seasonal storage. Raw stone blanks to make lance heads were also cached underground in hunting territories hundreds of miles from where the premium stone sources were mined or collected. These caches indicate access to vast areas of the Southwest and Southeastern United States among later Paleo-Indian and early Archaic groups at about 9000–7000 BC (Raff 2022), as well as the stocking of needed stone materials for tools and weapons (Kearns 2018). Thus, preplanned hunts are as obvious in the American Southwest as they were to the international archaeological teams documenting similar adaptations in Sonora and Chihuahua. As the Ice Age faded, major changes in climate and landscape triggered new technological, organizational, and logistical experiments throughout western North America and northern Mexico.

Heading South

Despite Clovis peoples' famed buffalo hunts, the most successful of early new world humans' cultural adaptations was the relentless southward spread of Pacific coastal Paleo-Indian peoples (Dixon 1999). In their boats and rafts they hugged the shorelines, subsisting on shellfish, fish, turtles, sea creatures, rich coastal estuaries, and a plethora of bird life. As the planet continued to warm up, their vast piles of discarded seashells created new, even richer coastline econiches. This was especially true at locations where freshwater creeks and rivers entered the sea and provided ample drinking water. Eventually, the massive shell middens generated by those early peoples trapped sand and created many new estuaries. In warmer climes, many of those estuaries eventually became vast mangrove lagoons, which are among the most diverse and productive of all human-exploited ecological niches. The mangrove econiches produced mullet, sea turtles, birds, bird eggs, crabs, conch, clams, oysters, sea urchins, kelp, otters, and duck. Those seashore habitats were rather similar, from California's Channel Islands to Tierra del Fuego's Beagle Channel. The notable similarities between these coastal ecozones sped up the process of human adaptations. For instance, kelp was harvested as a nutritious food from Alaska to Darwin's Beagle Channel.

Even without a mangrove cove, the seashore itself is a dramatically diverse ecotone. Ocean wave action continually interacts with beaches, rocky outcrops, cliffs, and shifting sand dunes, creating shallow pools that protected small fish and crustaceans from becoming instant pelican food. Such tidal pools dramatically enhanced local biotic diversity. In Mexico, heavy, ceaseless waves crashing against Sonora's basalt cliffs carved out water bowl-shaped pockets called *tinajas*. In the short rainy season, those seaside *tinajas* captured drinkable rainwater.

Ancient coastal peoples also made boats and canoes. Supported by favorable food abundance, they made their dash roughly eight to nine thousand years ago from the Alaskan coast to California and Sonora to the Straits of Magellan, then on to Tierra del Fuego Island and its Beagle Channel (Raff 2022). In stark contrast, early postglacial life in the northwestern *interior* of the North American continent was less generous because of lingering ice and snow. Thus, peoples of North America's continental interior scattered and changed slowly. Most of North America's northwestern interior was harder to negotiate, and the natural caloric yields of an intermittent creek-side camp or shallow, seasonal playa's sandy margins produced far fewer calories and a much lower diversity of edible foods than enjoyed by those who settled the Pacific coast. In short, population growth rates among interior groups of early Paleo-Indian peoples in ancient North America were significantly slower than among the Pacific Coastal groups.

The Paleo-Indian Southwest

During the early postglacial period in the greater Southwest, ancestors of later peoples were—like food resources—quite scattered. As postglacial temperatures rose, the surviving Ice Age animal species were reduced in both numbers and in body size. The most massive ancient bison species, *Bison antiquus*, had been replaced by a smaller species, *Bison bison*, by about 8000 BC. The long, heavy, stone-tipped lances of the late glacial/early Holocene Paleo-Indians had already been succeeded by lighter but longer lances. Even later, there came the atl-atl—an arm-extending notched board of three to four feet in length that hurled long, much thinner shafts tipped with skillfully worked fluted stone points. The atl-atl was probably a response to smaller, faster bison in growing herds nurtured by expanding grasslands as the early, wetter Holocene Period climate era wore on. The atl-atl darts reached velocities of nearly three hundred feet per second. The mass of the slender, flexible wooden shaft and its tip was multiplied by its velocity to create great penetrating power as its thin, sharp stone tip drove deep into the quarry. The cultural model of such postglacial animal hunts, often seasonal, was based on smaller, fleeter, and more numerous bison and other herd animals brought down by a group of organized hunters, followed immediately by efficient group butchering and meat drying on nearby bushes in the Southwest's hot, dry fall air.

These hunting strategies appear to have been in stark contrast to the earlier glacial era's patterns of hunting or scavenging one or two huge game animals. Deploying this earlier strategy, small hunting parties used their heavy, thick-tipped lances at short distances of ten to twenty yards to bring down a single mammoth or giant bison found mired in a muddy playa. Often enough, humans with lances also competed with wolves for part of their scavenged meat.

In short, postglacial hunting in the greater Southwest during a wetter period around 8000 BC displayed a hunting strategy that was, in and of itself, a temporary, culturally created, power phase. The temporary "killing and frenzied butchering" power phase was immediately followed by an efficient phase of drying meat strips on wooden racks or poles, then dry-air processing and transformation into jerky and/or pemmican. Pemmican included adding acidic chokecherries or juniper berries to the meat; the fruit's acidity tamped down on bacterial activity, preserving meat for winter diets.

The human work calories expended in the twenty-four to thirty-six hours of a bison hunt and its subsequent activities was huge—thousands of human metabolic work calories expended per hour were in stark contrast to an ordinary day-and-a-half of foraging or hunting small game. Yet the hunts proved efficient when multiple large animals were harvested. The trend toward larger, multifamily hunting groups during the Clovis era suggests cultural awareness of the huge benefits of multiple bison kills. Ironically, the rapid processing of butchering meat and hide consumed far more calories than the hunting/killing phase. Virtually all of the laborious hide preparation appears to have fallen on the women.

Note that, over the long haul, increasing climatological seasonality and enhanced migratory herd movements meant that large animals like bison were only available once or twice a year in a given plains locale. Bison herds moved south from the Rocky Mountains each fall, before snow packs trapped them, then migrated back north each spring toward the Canadian border. Thus, they passed through the central plains of Texas, New Mexico, Colorado, Nebraska, and Wyoming twice annually.

In contrast to yearly bison kills, the region's small groups of foragers who wandered and gathered scattered plant and animal resources expended far fewer calories per hour than did the bison hunters—but daily grass seed and rabbit foraging never produced a single day's caloric treasure to match the calories fixed in the meat and organs of a single bison or a desert sheep. It would take two or three foraging families several days to harvest enough wild dropseeds to feed everyone for a month or so.

These two distinct cultural patterns—hunting larger game versus foraging grass seed and hunting small animals—remained separated for millennia. This separation of food strategies exemplifies the specializations integral to emerging population growth and meat-focused tensions attributed to power phases and violent competition among hunter-gatherers (Stuart 1972).

For a time, these two food specializing patterns coexisted easily because the population groups did not usually compete for identical resources. In an ordinary winter, the big-game hunters and their winter meat supplies reduced fall season competition with plant foragers. In winters of dreadful snows, ones in which the bison were few, conflict over local resources likely ramped up. All that most small plant/small animal foraging groups had to do to avoid conflict was to stay out of late summer/fall big-game hunting areas. Yet, as climate changed to hotter and drier over the millennia, epochs of conflict were inevitable as millions of acres became dessicated. Life, for everyone, was much easier when there were fewer humans in a given resource district during periods of meager precipitation.

In the west, Clovis hunters seasonally hunted a species of very large bison (*Bison Antiquus*) across a huge swath of landscape as temporary multiband parties of hunters, which deployed to track and ambush the animals when bison herds appeared in their districts. Mid-fall to early winter seasonal hunts for bison were staged on the broken plains of New Mexico, West Texas, Arizona, and far northern Sonora (Kearns 2018). The peak of this brief cultural era spanned a century or two about 10800 BC. The Clovis Era was brief, dramatic, and widespread—a big game–based power phase lasting about three to four hundred years (Raff 2022), overlapping the earliest Archaic Period that followed.[4]

Seven hundred miles to the north in Colorado, Wyoming, Nebraska, and the Dakotas, the peak bison hunting seasons came in late spring and summer as herds moved north. As already noted, the herds stopped twice annually, making the central plains richer in huntable meat than elsewhere. In central Colorado, the Texas panhandle, and New Mexico's northeastern plains, the bison herds provided large quantities of protein-rich, fatty meat. Bison kills supported human health, pregnancies, and population growth. So did the fatty meat and nutrient-rich organs like liver, brains, and prized yellow bone marrow.

Lanced by spears and long spear-thrower atl-atl darts tipped with their distinctive fluted flint points, Clovis hunters and their immediate descendants, the Folsom people, left their technological lance-point calling cards and their efficiently butchered bone piles across a stunning swath of greater North America from Arizona to the mid-South. The expanding herds of downsized bison (*Bison bison*) that followed the disappearance of the ancient, giant elephant/mammoth species, then the later *Bison Antiquus* herds, coupled with new kinds of edible vegetation, had given these ancient bands of narrow spectrum hunter-foragers the gift of success. And as many modern Americans have recently discovered, bison meat is delicious, higher in protein, and far lower in fat than modern beef.

Neither the far north nor the far south terminus of bison migration routes offered as much hunting abundance and success as the middle region of the central plains, where the bison made their twice-yearly treks. In these districts, bison hunting began to geographically and culturally separate the most successful game hunters from family groups who pursued smaller, non-herding animals *and* foraged locally for plant protein.

Actually, these two lifeways were dramatically different in energetic terms. The foragers' daily food intake patters were predictable and constant: they worked almost every day and ate modest amounts of animal protein and fat over the course of a year. Their food consumption graph line, if charted, would appear rather flat. The most dangerous time of year for small-animal and plant foragers was the late winter to early spring. Foragers did hunt in fall, but their fall meat gathering focused on occasional deer and elk stalked in the upland forests. In contrast, hunter bands' food acquisition, if line-charted, would resemble a moving caterpillar. Storing jerked meat quite likely moderated the dietary dips in the chart during the hunters' late-winter hunger season.

In either group, individuals with the genetic tendency to store more body fat would have fared best, as would those whose genes rendered them metabolically more efficient. These two phenomena will be discussed in a later chapter.

More on the Clovis People and Climates

The ancient big-game and bison hunting Paleo-Indians are familiar to many modern southwesterners as "Clovis" people. Researchers have found distinctive and elegantly flaked thin lance heads in association with surprising quantities of bison bones at Blackwater Draw in New Mexico. Those finds are referred to worldwide as the Clovis type-site. The Clovis hunters reigned as an "apex" species, in direct competition with the wolf packs—one of the few remaining predators capable of bringing down a bison or elk, even in deep snow conditions (Stiger 2018). Hence, summer-fall hunts were the norm.

The original Clovis site is important in far broader context than most of its earlier investigators realized. These ancient Clovis peoples, descendants of even more ancient Ice Age hunters, bridged the dramatic succession of major climatological shifts that took place between roughly 12000 BP (Before Present) and 8,000 BP. They also spread rapidly in an arc across the southeastern US states, winding up in places as diverse as the mid-South and

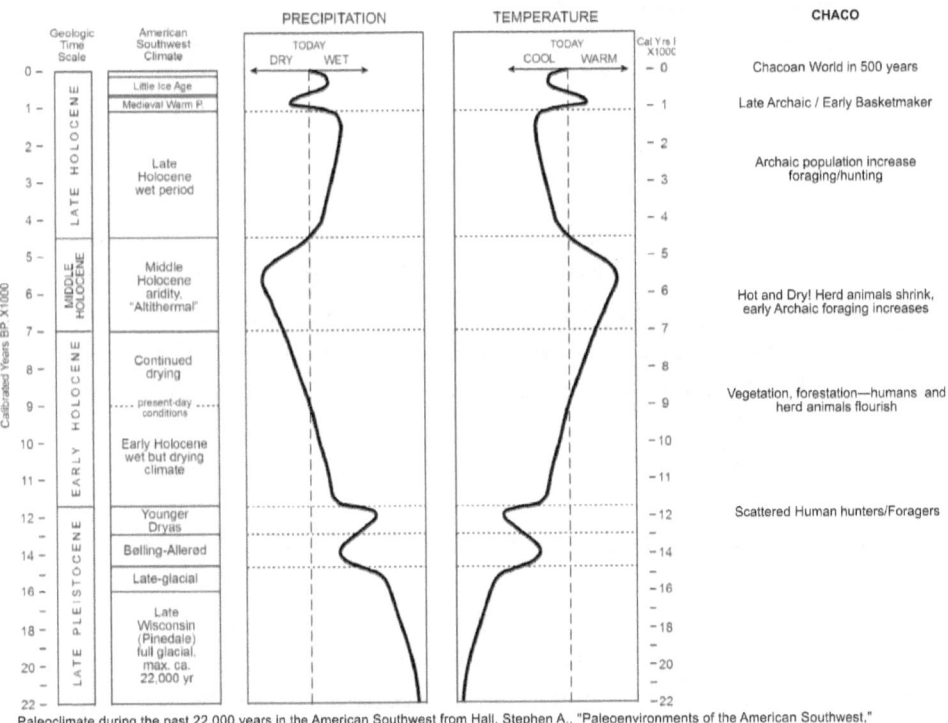

Figure 1. Paleoclimate during the past twenty-two thousand years in the American Southwest (this review); note scale change at twelve thousand years. Chart by Stephen A. Hall. Reprinted courtesy of the University of Utah Press.

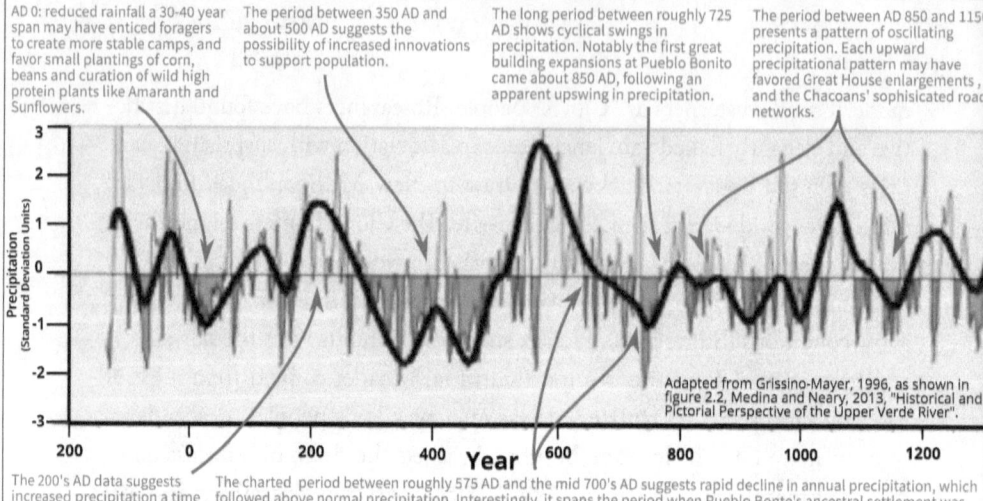

Figure 2. Estimated scope of Chacoan created water district about 1050 AD based on Morgan Hite's documented water flows and surveys done by author and Rory P. Gauthier in the late 1980s. Chart by Morgan Hite.

northeastern state of Pennsylvania. My friend and former University of New Mexico graduate school roommate, John B. Broster of Nashville, made a handsomely published career of meticulously excavating and documenting Clovis sites throughout the mid-South. So did Dennis J. Stanford, a legendary fellow classmate trained at the University of Wyoming and the University of New Mexico.

Though the climate charts above don't tell the full story, they allow us to understand that long-standing human adaptations to full glacial conditions became outdated as the northern hemisphere exited the Ice Age. It became "warmer and drier than full-glacial, but initially cooler and wetter than today" (Hall 2018). In short, the landscape slowly became mostly ice-free; sagebrush and grassy vegetation flourished, and many playa/lakes, some huge, dotted a wide swath of the Greater Southwest: northern Sonora, western Chihuahua, and what are now parts of southern California, Arizona, Colorado, and Utah. New Mexico was not left out. Three of its largest ancient lakes—Estancia, San Augustin, and White Sands—all have left us evidence of very early Paleo-Indian peoples.

Recently, human footprints have been found in the hardened sands of what was once a large lake in southern New Mexico's White Sands National Monument. Tentatively dated in news clips as twenty-two thousand years old, those footprints are likely more recent than that. Dating wet, rich sediment layers in ancient ponds and lakes is a tricky business, and a shallow lake that old was likely still frozen during the deep ice age.[5]

A Second Natural Power Phase

In thermodynamic terms, increased solar activity following the Ice Age combined with long cycles of planetary movements added heat calories to the northern hemisphere's atmosphere (Hall 2018). That heat caused a climatological power phase sufficient to melt ice, change the landscape, and heavily influence the kinds of vegetal and animal replacement species that had once populated prohibitively cold landscapes. The emerging postglacial landscapes demanded new human food acquisition behaviors, tools, and patterns of trekking movements to exploit increasingly seasonal and diverse plant and animal food sources. Given higher seasonal temperatures, the human dietary costs of simply staying warm had been reduced. As caloric body/warmth costs declined, nature allowed a larger share of consumed food calories to walk longer distances, reduce fewer cold nighttime calorie losses—shivering is calorically expensive—and redeploy those saved calories to enhance multiple work outputs.

In essence, Mother Nature had granted a warming climate that ended the Ice Age and bequeathed to the late Paleo-Indian era the solar calories needed to change ancient survival strategies, update toolkits, and successfully raise a few more children than in the more ancient, and colder, days of their forebears. These post–Ice Age hunter-foragers in the San Juan Basin lived about 8000–7000 BC (Hall 2018). They needed less firewood to ward off the cold and likely walked about five miles daily, an average distance based on ethnographic studies (Lee 1968), across warming landscapes. In the course of those walks, they realized an approximate 40 percent reduction in needed food calories when compared to their more ancient forebears, who had habitually walked through heavy snows. Decreasing firewood needs also allowed later Paleo-Indians the use of less densely forested districts, which harbored more small-game and edible plants.

Blackwater Draw and its Clovis type-site provide us the means to understand the consequences of yet another sudden climatic episode known as the Younger Dryas, a roughly twelve-hundred-year climate event that most specialists date to between roughly 12,900 and 11,700 BP. First came the rapid onset of a colder and wetter climate. Thus, spring brought heavier creek flows and extensive water-soaked meadows that created dense vegetation in soggy areas. As climate shifted again to much warmer and drier, parts of the Southwest's landscape again refashioned into sagebrush steppe. The welcome warmth of the postglacial climate diminished for a time during this period (Montgomery 2018). Subsequently the shorter, wetter, and cooler conditions of the Younger Dryas climatic event shifted to another era of warm, drying weather trends that enveloped the Southwest and was called the Early Holocene Period (Hall 2018).

If one takes the long view of these changes, they are dramatic. But if one focuses on the plausible average life spans of well-fed Paleo-Indian individuals as about twenty-five to thirty-five years, it would have become crucial for some band members to remember the seasons over a span of years, and to voice the temporary solutions to survival.[6] Elders were very valuable knowledge bearers.

From an ecological perspective, the results of the Younger Dryas's rapid climate shift were dramatic throughout the northern hemisphere. In northern Europe, average daily temperatures rose eighteen degrees Fahrenheit in just a month or so. In New Mexico, the rapid loss of lakes and marshy wetlands changed human lifestyles. New Mexico's Lake Estancia, White Sands, and the massive lake San Augustin all dried up over the course of a millennium. This altered bird, butterfly, and animal migration patterns. Change in the migratory patterns of bison herds and those herds' shrinking numbers interrupted many family bands' traditional seasonal patterns of movement and styles of hunting and foraging.

No single human generation bore all of the brunt of these changes in groundcover, temperature, lake (playa) shrinkage, and changes in plant and animal populations. Yet over the course of a number of generations, soils dried out and frequent dry winds began to scour the Southwestern landscape and its Paleo-Indian peoples. By the end of the Paleo-Indian period, roughly 6000 BC, the landscape was desiccating across nearly all of the greater Southwest. This is evidenced by hand-dug wells at Blackwater Draw and similar landscapes across much of the lower elevations of the American Southwest.

Another line of fascinating evidence has been produced by archaeological work in the Mexican state of Sonora. It is roughly dated between 5000 BC and 2500 BC (Hall 2018), a period when rising heat made water scarce—even the Great Lakes far to the north partially dried up—and soils became much more alkaline. Those factors would radically change grasslands and plant species, forcing Paleo-Indian peoples in northern Mexico to again readapt to their environment.

The Gomphothere Elephants

As has been established, Clovis Paleo-Indians had begun to inhabit Sonora and western postglacial Chihuahua about 11,550 years ago. Recently published research by John Carpenter, Guadalupe Sanchez, and Ismael Sanchez suggests that the Clovis bands arrived along the Pacific coastal route as their descendants moved ever southward. That scenario, and the precise dates, are still debated. Yet, the archaeological data indicate that, in Sonora, these groups traveled wide distances annually. Their largest camps focused on places where groundwater, flint, and/or game were plentiful. To quote Carpenter, Sanchez, and Sanchez,

> Sites known as "Fin del Mundo," "El Bajio" and "El Aigame" represent various Clovis family groups that utilized local raw material to make tools. The assemblages make it evident these groups shared tools between them, suggesting close family/group relationships (Carpenter, Sanchez, and Sanchez 2018).

These Clovis data from northern Sonora are quite eye-opening. The very ancient dates of Sonora's Clovis tool complex indicate that the earliest Clovis tools there date to about 11,500 years ago—a time when Sonorans hunted small, straight-tusked elephants of the ancient *gomphothere* genus. The Sonoran gomphotheres were among the last of the elephant species in North America.[7]

Some scholars believe that Clovis may not have originated in the high plains of Wyoming or Colorado as long thought (Carpenter, Sanchez, and Sanchez 2018). The Clovis adaptation may have first been refined in far southern California or Sonora and northwestern Chihuahua, then worked its way north to cooler places like New Mexico's Blackwater Draw and the west Texas Clovis sites.[8] Areas to the north and east of Sonora may have offered ancient hunters a temporary climate buffer as the Younger Dryas rendered Sonora too hot and hostile for a big-game focused culture. There was too little water to nourish large, semidesert brush and trees, the primary food of the elephantoids.

It is likely that successive waves of former gomphothere hunting Paleo-Indian populations in Arizona, New Mexico, and West Texas collided with groups moving north from the rapidly desiccating plains of Sonora. Such a collision of in-migrations and rising human population density could easily have triggered cultural change and higher levels of food competition. This may have induced new dietary experimentation with foraged plants, grass seeds, and small game.

The Altithermal Period: Northern Mexico's Changing Landscape in the Altithermal Period

By about ten thousand to eleven thousand years ago, the Early Holocene (modern) climatic regime brought landscape changes in northern Mexico, a wide swath of the greater Southwest, and high plains to the north. These changes transformed wide areas of northern Mexico and the adjacent United States into a simile of Sonoran desert landscape. Juniper and oak were partially replaced by mesquite, palo verde, cactus, annuals, and other arid land grasses. The large Pleistocene animals were gone, replaced by smaller, but far more numerous, species: birds, rabbits, smallish deer, desert sheep, wolves, foxes, coyotes, pack rats, wood rats, turtles, tortoises, snakes, lizards, and prairie dogs. Yes, some bison herds survived—but even they had been downsized by Mother Nature and are known to zoologists as *Bison bison*, which, as noted earlier, replaced the larger *Bison Antiquus*.

Agave sketch by Esther Burton.

Also noted earlier, the regional archaeological/artifact inventories suggest Sonorans were moving north into what is now the western United States (southern California, Arizona, New Mexico, Texas) at roughly the same time as some far northern Paleo-Indian peoples from Canada and the Dakotas were beginning to move south. Those regional movements increased population density from Wyoming to the plains of southeastern New Mexico. Rising human population density suggests the likelihood of rising competition for food resources.

As seasonality became more pronounced, rising temperatures again dumped much higher levels of solar energy into the Southwest's landscape during a climatic event called the Altithermal. Daily temperatures rose quickly and dramatically in the northern hemisphere—data from far northern European ice cores indicate a rise of ten to fifteen degrees Fahrenheit in about two weeks! In the American Southwest, these changes occurred between roughly 5000 BC (7000 BP) and 3400 BC (5400 BP). Dramatic heat waves created sere landscapes.

Suddenly, the Altithermal landscape in the Southwest transformed the lowlands from lush grasslands, oak and piñón forests, and large lakes to a world of yucca, short grass, and cactus. The large bison herds diminished over several centuries, rendering Clovis-like, big-game adaptations to the category of wistful campfire legends of giant game and tall, well-fed hunters. Heat, rapidly drying water sources, and increasingly thin, alkaline soils forced the changes to a hardscrabble way of life known as the Early Archaic. This emerging Archaic way of life relied on a much broader spectrum of plant foraging and smaller animal hunting than among the vanishing Clovis cultures. Plant and small animal foraging had, out of necessity, become cemented in the cultural behaviors among most of the regional populations that survived the Altithermal.

Of great evolutionary import, the demographically most successful human groups were those that re-strategized to efficiently map themselves onto rapidly transforming landscapes and increase their knowledge of new animal and plant characteristics. Instead of targeting large-bodied animals that were few in number, this era dictated the pursuit of smaller, more plentiful plants and animals like rabbits, whose sheer numbers allowed their breed to stay alive despite the Altithermal's hard drought years. Smaller, shorter-lived creatures like rabbits required less caloric sustenance; thus, they often produced large and frequent litters.

Amid these climatic changes and transformations in lifestyle, it is likely that adult human height and weight declined during the Altithermal (Fogel 2012). This assumption is based on growing population, episodes of reduced forageable groundcover, and lower intake of protein than in the antecedent hunter-gatherer populations. In modern human populations, the result of poor nutrition is statistically obvious when the next generation of children born is shorter and sicker. During the great famine years of the 1840s in Ireland, the average adult height of surviving children dropped about two inches in one generation. In the United States between 1830 and 1860, a 1.1 inch decline in male height occurred (Steckel 2002). That decline was a cofactor of social unrest, rising food prices, and static wages, as well as a signal of economic desperation.

Cultural Trends and Climate in the San Juan Basin

The changing climatic trends in the Southwest that began about ninety-five hundred years ago and ended in the Altithermal appear to have profoundly reshaped the region's earlier human lifeways by roughly eight thousand years ago. As the Altithermal's heat and aridity softened (about sixty-five hundred years ago), broader-spectrum foragers gathered, stored, and hunted a wider variety of species than previously. Food storage in dugout jar-shaped caches multiplied at seasonal camps, and plant processing became more efficient. These trends set the Archaic Period on a path that led to increasing population density in the region surrounding New Mexico's San Juan Basin. Food stability encouraged increasing experimentation with food processing, more efficient gathering tools, nets made from human hair or yucca fiber, early grass stem basketry, and more efficient cooking techniques. A flat rock laid atop a small fire in a shallow pit allowed griddling all manner of ground grass seeds mixed with a few berries, *chenopodia* leaves, yucca buds, and a touch of meat—lizard, gingko, rabbit, ground squirrel, or even grasshoppers or cicadas—were consumed. Mice, wood rats, box turtles, and birds' eggs might all be mixed and cooked. Some small seed grains like Indian ricegrass were likely ground fine and consumed in water. We know this as *atole*.[9]

Territorial Tensions

As the glory days of Paleo-Indian big-game hunter culture faded in the protracted heat and aridity of the long, harsh Altithermal Period (7000 to 4500 BC), it is important to understand that small groups of remaining hunters were not seen as primitive by foragers but as cultural icons of a golden age when giant men armed with huge atl-atls ruled the Southwest's hunting fields. They were few in number, fiercely guarded their territory, and likely had several wives.

By the dawn of the Archaic Period in the San Juan Basin, roughly dated to 7000 BC (Kearns 2018), descendants of hunting-focused peoples had occupied the region since the waning of the Ice Age. Of course, tales of these hunting giants and their traditions were deeply rooted in legend. Those who lived on to roam the drying landscape in search of game during the harsh, population-limiting period of the Altithermal climate seven thousand to forty-five hundred years ago would have left dart/lance points in various ancient styles as they scoured for game along ancient lake margins rendered dusty and sandy by the Altithermal (Hall 2018).

Legends and fantasies of hunting grandeur died hard in successive foraging societies. The mere size of those Paleo-Indian points sent a heady cultural message across the ages: "The giant hunters were here. They were our forefathers." Thus, the remaining scattered hunter groups persisted and, for a time, dominated the Four Corners district, long after large-game hunting had begun to fade. The rub, of course, came when foragers began to intrude on hunting lands as they harvested grass seeds, small game, pine nuts, and acorns on the fringes of hunter territory. Even as young forager men hunted rabbits, their mere presence in the big-game territory of mountains and forest led to conflict, some of which has been documented as peaking in the regions northwest of Mesa Verde in the 700s to 500s BC. That is the region where the atl-atl was first replaced by bow and arrow in the Southwest (more on this later), as two competing ways of life coexisted: ancient and iconic hunting versus new and more efficient plant and small animal foraging.

Rising population in the San Juan Basin led to increasing competition for territory. Nothing was guaranteed, unless one could create one's own Garden of Eden. To that end, we will see that the first diverse, female-created gardens began as necessary experiments with cultigens and foraged seeds. Occasional gardening in the San Juan Basin at about 5000–4000 BC morphed into several distinct garden styles by about 1000–1500 BC. All of what became the Chacoan World from about 500 AD to 1200 AD had come to depend on the genius of broad-spectrum gardening, which was usually women's work. That legacy of female gardening would survive well beyond Chaco Canyon's glory days. The day would come when a young Pueblo woman in love would prove her marriage worth by using her gardening and cooking skills to grow and prepare nutritious food, a role so important that it would become a cultural legacy and the probable basis of Puebloan female land ownership. For a time, the women's garden food production enhanced female status in the emerging Basketmaker world that would follow.

CHAPTER 2

CHANGES IN THE FOUR CORNERS REGION DURING THE EARLY ARCHAIC PERIOD

SIGNIFICANT CHANGES IN TECHNOLOGY, landscape, climate, foraging patterns, and in foods eaten announced the onset of the Early Archaic Period, circa 6800–3600 BC (Kearns 2018), in New Mexico's San Juan Basin. This rugged territory would, thousands of years later, produce the complex society that scholars have labeled "The Chaco Phenomenon." It is hard for most of us to grasp such lengths of time, for our modern world is, if anything, pathologically fast-paced. We live fast, forget just as rapidly, and indulge in awesome levels of waste.

In contrast, archaic time was measured by moon phases and seasons. The moon offered clear time frames, as did solstice sunrise over special horizon points, but history was archived in the depths of the living human brain folds of the elders. Memory was history. It functioned as the Archaic world's database. Elders with good memories were revered.

Early Archaic peoples lived rather different lives than had their Paleo-Indian forebears. The bulk of food calories consumed by them shifted from hunted meat to a mix of less meat and more vegetal foods. Food was where one could find it, learn to process it, create ways to store it and, as regional populations grew, to define accessible territories from which to collect it.

Nature itself offered a multitude of seasonal clues: the arrival of spring birds; the fall flights of migratory geese overhead; the annual succession of passing butterfly species, newly born rabbits, the subtle appearance, or sudden disappearance, of particular hummingbird species. The skies also offered daily clues about weather. These clues, and even many more subtle ones—like the movement of ants, the activity of termites, a frenzy of squirrels harvesting hillside nuts earlier than usual—were all noted and processed by family groups struggling to survive ever-changing climatic episodes, while avoiding areas of increasingly dense populations. During the Archaic Period, the daily world required higher levels of food production to support mobile foraging families, which meant more walking and more work gathering food.

Archaeological data from multiple sources suggest that during the Early Archaic Period, foragers ranged widely—over hundreds of square miles in the course of an annual cycle. As best we know, those annual cycles of movement drew people to the uplands in early spring, then to the lowlands in summer (Kearns 2018). Fall camps tended to be found in higher elevations. Winter camps could be low or high in elevation; the latter depended on a warm updraft in a given canyon head, a pronounced southwestern exposure, or an east-facing hillside to capture the heat of the sunrise.

After the leanest months of late winter, hungry family groups deployed in early spring to harvest the first crops of abundant seeds of Indian ricegrass and other bunchgrass varieties. These protein-rich seeds nourished humans, most of whom had suffered weight loss after a winter camp's jerked meat ran out and winter snows inhibited organized hunting parties.

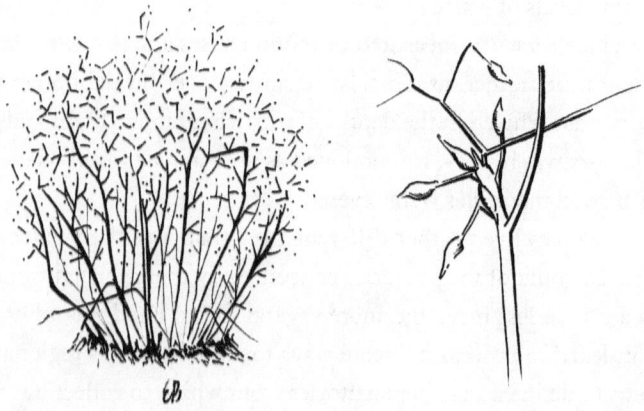

Indian ricegrass sketch by Esther Burton.

In years of unusually deep snows, coyotes, wolves, and wolverines easily negotiated the snowdrifts, but humans and most large game animals were stymied. Groups of deer often "yarded up" during a blizzard, packing together to conserve body heat; in that process, their collective body heat sometimes melted the snow around them and left them trapped in an ice pen created by their instinctive attempt to survive. Most often, those frozen deer were consumed by wolves, wolverines, bobcats, and coyotes. Occasionally, in a stroke of luck, a winter hunting group would have beaten the predators to the treasure trove of venison, fat, highly nutritious long-bone marrow, and hides. While some early family groups continued to favor large game hunting, the majority had begun to invest in hunting small animals and foraging a wide variety of plant species (Kearns 2018).

According to Timothy Kearns's meticulous analysis of recorded archaeological sites in the San Juan Basin, it appears that on the cusp of the transition from the Middle to Late Archaic Period, recorded Archaic sites declined in number. That transitional time frame spanned 3300 BC to about 1500 BC when recorded archaeological sites dipped sharply in numbers. This decline in site numbers could imply larger foraging group camps; alternatively, it may indicate a period when food competition became contentious enough to have pushed some populations out of the San Juan Basin. Either scenario, or a combination thereof, is possible (Kearns 2018).

Over time, use of the landscape had shifted. In the Early to Middle Archaic, many ephemeral camps were created. A number of them are found on the open basin floor. Hand-dug storage pits were sealed over and left, filled, for use upon return. This pattern does not suggest extreme competition, but the reduction in numbers of sites in the Middle/Late Archaic Period about thirty-two hundred to thirty-eight hundred years ago is notable, and requires more research.[1]

As the Middle Archaic's hot, harsh Altithermal climate faded, archaeological evidence indicates that many more family groups turned to broader spectrum foraging. This makes perfect practical sense—cooler temperatures and higher levels of annual precipitation would have rapidly expanded the range of edible plant groundcovers. A combination of rain/snow water and moderate temperatures encourages grasses—sand dropseeds, various grama, Indian ricegrass, bunchgrass, buffalo grass, and the like. In practical terms,

TABLE 1. Archaic Periods in the San Juan Basin and Thermodynamic Trends

1. Early Archaic	6800 BC–3600 BC	Broad-scale foraging and game (medium)
2. Middle Archaic	3300 BC–1500 BC	Broad-scale foraging; population increase; some game
3. Late Archaic	1300 BC–500 BC	Foraging and small-scale gardening; less meat hunting
4. Basketmaker II	300 BC–500 AD	Settled homesteads; gardening, foraging, small to medium game on occasion

Adapted after Timothy Kearns's data (Kearns 2018)

TABLE 2. Common Seasonal Wild Plant Foods

Indian dropseed and ricegrass	Acorns
Amaranth	Wolfberry
Sand dropseed	Chokecherry
Grama grass	Globemallow
Sheepsfoot, goosefoot, other chenopods	Buckwheat
Cactus, *nopal, cholla, agave*	Potatoes
Piñón nuts	Sunflower seeds

the expansion of early spring seed crops of plants like Indian ricegrass encouraged changes in diet, logistical foraging behaviors, and cyclical foraging movements across the landscape. Once foraged, most plant foods needed to be processed and, when and where surpluses were possible, stored in some fashion.

The spring staple Indian ricegrass does not shed its seeds in neat piles. If unharvested, the seeds that drop spontaneously, or are dislodged by grazing elk and deer, blow away to be eaten by many other small animals. Thus, those seeds need to be carefully hand-stripped from the plant's curled stems. This is where the coil of braided grass stems, mentioned in the introduction, comes into play: coiled grass baskets held the precious seeds stripped from sloping ricegrass fields by an early spring foraging party of perhaps a half-dozen people.

Goosefoot sketch by Esther Burton.

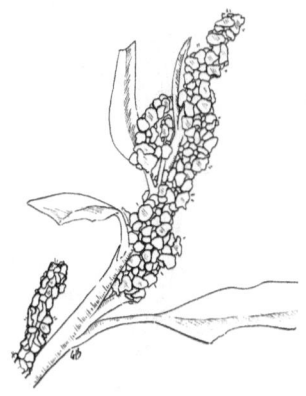

This seed collection was essential for both immediate use and fall storage to reduce late-winter human weight loss. A foraging group could not simply gorge it all and also expect to grind and store it for fall-winter camp supplies. Unlike hunting groups, which tend to binge eat, seed collectors eat moderately and more often than hunters. Late winter and early spring were the seasons of scarce food, weight loss, vitamin deficiency, metabolic syndromes, failed pregnancies, and empty storage pits.

Midsummer was another challenging food season, as peak temperatures often combined with cycles of brief, spotty rains. The heat and aridity wilted many nutritious, leafy plants, among them members of the *chenopodia* plant family, like goosefoot, which is still cooked by both Native Americans and northern New Mexico *hispanos*, who know it as *quelites*. Quite nutritious and iron-rich, it was eaten much as many now eat spinach.

During the more densely populated Late Archaic Period, spring was often a season of scavenging old growth: Shriveled berries that birds had overlooked, as well as a rabbit here, a gecko there, a box turtle in hiding, a prairie dog or a rattlesnake carefully pulled out of a rocky crevice—all provided meat, organ fat, and protein. Circling turkey buzzards gave reliable clues to the location of larger animal deaths. If reached quickly, some meat, hide, and valuable bone marrow might be procured. One could not be picky about eating fly-blown meat in the hard times of summer droughts. Animal bones also provided valuable material for tool making.

When monsoons did rehydrate the landscape in July, August, September, and October, vitamin and protein-rich wild amaranth, chokecherries, wolfberry, and piñón offered the chance of full storage dugouts by late fall.[2] Favorite Middle Archaic winter campsites were not on the valley floors during this period (too cold)—rather, they were typically located in the hilly piñón-juniper ecozones ideal for fall camps near large game territory. That game was typically a deer bedded down in the deep hillside

Wolfberry sketch by Esther Burton.

thickets at about sixty-five hundred to seventy-five hundred feet in elevation.

These winter camps were often found on hillsides facing south, east, or west. Few faced north—too cold in winter. No one lived above the tree line. Winter temperatures were lowest in the valley bottoms and in the high mountain forest zones of ponderosa and aspen. Exposed, south-facing, elevated hillsides were three or four degrees warmer than either of these areas. Even now, on a late winter afternoon, one can drive east up Albuquerque's Central Avenue from the Rio Grande to the heights near the University of New Mexico's Nob Hill district and watch the dashboard thermometer rise from thirty degrees near the river to thirty-four degrees in Nob Hill.

By about 2000–1500 BC, those ancient Archaic hillside winter camps often included one or two shallow, brush-roofed, sleeping dugouts, charcoal-rich hearth areas, fire-reddened stone, a dugout jar or bell-shaped food cache, and a quantity of ground and flaked stone tools.[3] The comparatively few sites in the Four Corners region where sample excavations have been conducted yielded small bone fragments and discarded bone tools like awls, needles, and small scrapers.

Few of these Archaic Period forager sites have been fully analyzed by advanced laboratory methods. The costs of such analysis are generally prohibitive given ordinary archaeological survey budgets. In typical soil and site detritus samples, researchers find a wide variety of plant seeds or their husks. These include sunflower, Indian ricegrass, goosefoot, tansy mustard, common dropseed, and wild amaranth (Montgomery 2018). Other food remains often found include yucca, purslane, piñón nuts,

Tansy mustard sketch by Esther Burton.

WHERE DID THAT FOOD COME FROM?

In spite of cost impediments, it would still be worth funding studies of the microscopic differences in ecotonal varieties and sub-species of ricegrass found in storage pits or among the detritus of ancient shelter floors. The ricegrass species/subspecies varied subtly by ecotone. Varieties of ricegrass that flourish in the higher elevation piñón-juniper ecotone are often *not* the same species or varieties that prosper in lower elevations, in salt desert scrub ecozones and sagebrush steppe plant communities (Oyle 2013). If even a few dozen seed samples could be analyzed, scholars might benefit from the answer to the question, "Where was this foraged, and where was it stored?"

various chenopod species, walnut, oak/acorns, and various cacti fibers and seed pods.

Late Archaic-era animal remains include jackrabbits, cottontail, prairie dogs, pack rats, gophers, occasional deer, desert sheep, or pronghorn bones. The smallest animals, like geckos and other lizards, were eaten but often wound up consumed as part of a mash mixture of seeds, wild edible leaves, berries, and essence of pulped tree or fence ginkgo. In the "Ancient New Mexico" classes I taught at the University of New Mexico for more than forty years, I labeled these mashes as "critter fritters." A flattish fire-heated cobble the size of a doorstop did the toasting work. While such a food now seems exotic, it was, in fact, quite a sensible broad-spectrum meal: plant and animal protein; a range of vitamins and essential minerals like magnesium and iron; a touch of sugar from edible plant stems, cactus, or seasonal berries; and plant fiber. Ancient critter fritters offered fewer calories but much more balanced nutrition than today's "quick food" drive-ins. Note to those who view ancient Native Americans as "quaint" or "primitive": Many ate better balanced diets than most modern Americans. It was quantity more than quality that challenged the Archaic era's occupants of the San Juan Basin. When the spotty summer rains did not come, humans suffered and lost body weight.

Improving nutrition redounded more to broad-spectrum foragers/gatherers than to hunter-focused groups. In years of ample precipitation, foragers would not have suffered the same level of late winter/early spring vitamin-deficient metabolic syndromes as did the hunters. Big-game hunters tended to binge eat. When their jerked meat ran out, hunger followed. Plant foragers, however, were more likely to eat smaller, more regular meals; that habit coupled with careful food storage was crucial to their survival

Much climatic and ecological data came from the pack-rat middens excavated in the greater Southwest. Pack-rat midden analysis has resulted in a forty-thousand-year record of climate and vegetation changes (Betancourt, Vandevender, and Martin 1990). Pack rats average about fifteen ounces in weight, and made a good base for a nutritious critter fritter; that is, if bobcats, coyotes, or wolves did not get to them first. Females produced litters about every six to eight weeks, ensuring biological replacements.

Analysis of ancient pack-rat middens has given modern archaeologists and ecologists valuable clues to general climatic conditions and edible plant species at different time periods in the Four Corners region. Pack rats survived every twist and turn of ancient, often radical, climate events. They lived solitary lives in caves, canyon walls, rock slits, and emerged to forage only at night. By foraging juniper berries, seeds, and edible twigs within a hundred yards or so of their midden nests, they lined their "larders" with regional plant samples that could be radiocarbon dated. They survived droughts easily, as they drank no water. Rather, their bodies absorbed water from their consumed plant foods. They then excreted a thick, moist pine smelling slush known as amberat, which hardened into a brick-like consistency (Betancourt, Vandevender, and Martin 1990).

Work Roles of Hunters Versus Foragers

Broad-spectrum foraging work was more inclusive than big-game hunting—virtually everyone from the age of five to sixty-five could do it. Thus, the labor pool was broad and robust among foraging groups. Processing the harvested wild seeds and weaving baskets to carry or store plant seeds took skill. Based on widely cited ethnographic trends, most of those grass weavers were likely women.

Social Dynamics of Hunters

The large-game hunting bands awarded high status and preferential sexual access to the best and most experienced hunters and their protégés (Lee and DeVore 1968).[4] Female status, as recorded in the extensive ethnographic literature on hunter-gatherers, would have been *much* lower than that of the dominant males.

Hunt seasons, their staging, scheduling, strategies, and targeted pursuit locations were all decided by the most successful male hunters. Successful hunters were likely those who had been best fed in infancy and childhood. Proper nutrition would have allowed their physical frames to be taller with longer arms, which enhanced the distance and velocity of their atl-atls' three-to-four-foot-long "darts."

A single high-status father with access to ample meat and well-fed female sexual partners could potentially father all of the next generation's elite

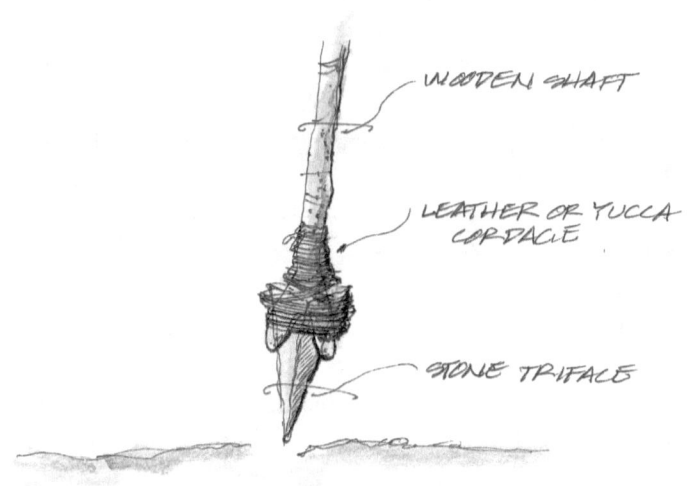

Hafted digging stick with triface stone point (Manzano Mountains). Drawing by Baker Morrow.

hunters.⁵ The young, short, or weak males were often out of luck socially and lived relatively impoverished sexual lives.

During a mid-fall hunt, the men would have departed camp, leaving behind the women, children, and likely an elderly bowman or two. Men whose vitality or eyesight was fading also stayed behind. Hunting depended on the cultlike skills of a few able men. Atl-atl skills took several years to develop, so young boys could not immediately go with their fathers or uncles on the hunt. While the men were away on a long hunt, women and children would forage and hunt locally with throwing sticks or shorter atl-atls.

Depending on the target species, the deployed hunt party might have

ANCIENT LINEAGE OF DOGS

Domesticated dogs had ancient origins leading back into the last Ice Age. By the time the first humans migrated to North America from Beringia and far northeastern Russia about thirteen to fifteen thousand years ago, they brought with them dogs that had already been domesticated. Their formative and ancestral populations—wolves mixed with a genetic dash of Asian golden jackals—had died out before the San Juan Basin was populated, leaving only fragile traces of diluted DNA. Thus, those ancient peoples who first migrated to Alaska from Siberia twelve or thirteen thousand years ago brought dogs who did not survive to become the future gene pool for later North American dogs.

Currently, their genetically closest living modern dogs are Alaskan breeds: Huskies, Makenzie River dogs, Malamutes, etc. These dogs still howl like wolves and have the same eerie, spatial awareness that wolves have. The female husky puppy Nokotch ("Big Eyes" in the Tanaina language), who left Alaska on my lap in a float plane at the end of the 1969 archaeological season, was the first purebred husky to reside in Albuquerque.⁶ She watched airplanes endlessly, following their trajectories in the sky above the city.

brought along dogs. If deer or desert sheep were the quarry, the dogs would get the first scent and drive a lone mule or whitetail deer from its bedding in deep brush. If a high-country bighorn ram or mountain lion was the target, dogs were less useful, as they could not negotiate the rocky precipices.

If a fall hunt focused on an isolated bison, the dogs would be held back, and hunters would maneuver onto hillsides to prepare their assault. Sited above the straggling bison, their distinctive human sounds and scents would drift away far above the unsuspecting animals.

If a hunt was successful and relatively near the group's temporary base camp, the women, children, and elders would deploy to assist in the butchering. A single deer would be rough butchered, the meat divided into "packages," and the hide shouldered back to camp by several men. The game tracking and killing was a cultural right owned by high-status males; the processing, cooking, drying/jerking, and hide preparation were the "lower-status" obligations of the women. Note that the preparation of a hide was a *very* arduous multistage task.

Both the cult of maleness and the cultural presumption of that gender's high status must have nurtured a male-focused, skills-tested ethos: "I don't do grass seeds . . . that's women's work." In modern times, that lingering sense of innate male status can still produce a husband/male partner who eschews unglamorous household tasks. Women typically worked longer—then, as now.

The hunting way of life did have benefits: it endured for many millennia under low population densities. It likely produced slightly taller people who benefited from the nutritional outcome of a relatively higher-protein and fat-laden diet.

The male ethos, however, *much* favored sons over daughters. Males consisted of over 50 percent of the overall population. This gender bias undoubtedly produced fewer children. If fraternal male and female infant twins were born in a food-poor late spring, the likelihood was that only the male infant would be nursed. The infant female was more likely to be abandoned, smothered, or refused a breast, unless another woman in the group could be a wet nurse, perhaps due to the loss of her own late-term fetus or stillborn child. This would limit a band's size.

Social Dynamics of Foragers

Plant forager groups' gene pools grew in diversity over the centuries because, unlike in hunting society, no small group of elite males commanded sexual access to the majority of the young females. That led to more females able to bear children *and* more valuable genetic variability than in a population dominated by only a few male Y-DNA gene types.[7]

In addition, newborns were less likely to be smothered or abandoned, and dominant males' lack of sexual access to younger women also would have been muted as compared to the hunters. Thus, there were simply more females in the forager bands, and that likely generated a more normal sex ratio nearing a percentage of fifty-fifty, male to female.[8]

Broad-spectrum foragers also appear to have been more egalitarian in their labor practices. Ethnographic data tell us that women were typically the masters of plant foraging and processing. Five-year-old children of both genders began to forage effectively under close female supervision while the men slipped away to hunt locally.

The lower elevations of piñón-juniper ecotonal lands were one of the foraging bands' prime territories. If potable water was available at a distance of a half mile or less, so much the better. Nearby brushy hillsides were prime locations for seasonal camps. Here, both cottontail rabbits and jackrabbits could be hunted. The jackrabbits gravitated to dense stands of tall brush; the cottontails favored vegetational edges like paths and field perimeters. Deer often bedded down in the brushier piñón-juniper stands or deep ricegrass pastures near these camps. Thus, a camp overlooking a grassy valley was ideal. Forager camps typically consisted of multiple hearth areas, suggesting related multifamily groups living together.

Food Efficiencies of Hunters

Hunter groups saw the landscape as a series of shifting targets. They sought exclusive, long-term access to the territories most likely to produce large game. These hunting groups did forage plant foods when survival was at stake, but their culturally ordained food targets came in the form of high-calorie, high-fat, and protein-rich meat. Their larger bodies, courtesy of meat

STUDYING FORAGER CAMPSITES

Women foragers' campsites left behind soil enriched by human excrement, dropped grass seeds, small animal bones, a few broken tools or grinding stones, and even the plant fiber wads women used during menstruation (W. W. Dunmire 1995).

The disturbed or damp soils of these seasonal forage camps nourished plant species like amaranth and the iron-rich leafy *chenopodia* family. After the hot Altithermal Period ended and winter temperatures became colder, these dwellings morphed into deeper pithouses. Seed grinders, often round basaltic stones, can now be found near many ancient hearth areas. Small tools like awls and scrapers fashioned from either flaked stone or the animal bones of small birds, quail, wild turkey, rabbit, or deer are also found in campsites of this era.

To properly excavate and evaluate such a foraging site takes both time and a high level of archaeological laboratory skills, but most archaeological projects are limited due to stingy resources. In spite of that, the last three generations of Southwestern archaeologists have done an amazing job of teasing out many details of the broad-spectrum forager strategies. In my view, the best of the best laboratory reports on late Paleo-Indian and early broad-spectrum foraging sites have been written by UNM scholars Drs. Bruce and Lisa Huckell, as well as by the legendary Dr. Richard Ford.

---

and fat consumption, required more annual dietary calories per person than among the plant and small-animal foragers—large size and robust muscle mass are calorically expensive.

Long-term and restricted access to their favorite staging and hunting territories was the hunters' main advantage. Having an annual fat and protein intake that was almost assuredly higher than that of their broad-spectrum plant foraging neighbors was another.

In terms of efficiency, efforts in saving and storing food were intermittent.

While hunters' tendency to binge eat initially enhanced a hunter's excess body mass, overall health would steadily decline during starvation season. "Self-storage" by means of binge eating could only work if a hunting band selectively protected the most successful hunters.

Food storage was limited in the hunter groups' campgrounds. Hunting women foraged, but they could not save as much gathered food as those living in the seasonal, broad-spectrum foraging camps at lower elevations. That pattern led to occasional food shortages among hunters, an inefficiency that raised the specter of hungry hunters who probably engaged in raiding broad-spectrum forager camps when they were at risk of starvation.

Jerked meat and pemmican were probably the hunting groups' best "efficiency" practice. Pemmican was made after a harvested animal's organs were consumed. Those portions of a hunting group's meat not eaten in the first few days at a kill site's camp were typically cut into long strips, then strung on drying racks or nearby bush tops. Smaller pieces of meat were often pulverized, then dried after mashed berries and seeds were added.

Hunting society favored competition and exclusion. The biological and evolutionary outcomes of these behaviors led to a culture that was forced to limit population growth. Over the next twenty to thirty human generations, a span ranging six hundred to nine hundred years, the ancient hunters' reproductive conservatism and their asymmetrically high male-to-female sex ratio would gradually render them an ever-smaller percentage of the Four Corners' population.

Food Efficiencies of Foragers

While hunting group populations were forced to limit population rates, broad-spectrum forager populations continued to grow freely in numbers. An increase in the size and numbers of their campgrounds' bell-shaped storage pits, hearths, and shallow dugout shelters confirm this upward growth trend.

In sharp contrast to the hunting groups, foragers sought hard to access plant-based protein. Piñón nuts and acorns rich in protein could be gathered and processed in the high country. Foraging males did hunt larger game, but it had likely trended to opportunistic.

Changes in the Four Corners Region During the Early Archaic Period

Piñón seed, cone, and tree sketches by Esther Burton.

Foraging required cooperation. This alone was responsible for the rise in forager populations as they demonstrated their ability at finding numerous food targets, organizing camps, collaborating on work roles, and cleverly storing and using what they collected.

Base camps served as anchors that melded foraging families and landscapes into efficient, productive, and diverse ecosystems. Robust seed-based food storage supported the growing broad-spectrum forager population. The stored rations undoubtedly saved some infants and winter pregnancies each year. In addition, foragers improved their general health by avoiding binge eating.

The Role of the Turkey

For both hunting and foraging groups, small-animal meat—gophers, pack rats, snow rabbits, lizards, voles, box turtles, and the like—would have diminished

Acorn sketch by Esther Burton.

PROCESSING ACORNS

Processing acorns required leaching the extracted plant meat with fresh water (to reduce the acorn's bitter alkaloids), then grinding the meal with one-handed *manos*, round stones used to pulverize acorn mash on a larger base stone called a *metate*. This was laborious work. The process was most efficient when either flowing creeks or a quantity of melting snow runoff was available; thus, hunter-forager groups likely foraged acorns in the late winter months. The rich protein and abundant vitamins in acorns, however, rendered the tedious process worthwhile. The acorns were then ground into a nutritious, high-protein flour that could be added to stew pots or formed into small cakes mixed with dried, sweet berries. Acorn meal was also acidic enough to reduce spoilage of many dried foods.

the seasonal metabolic syndromes like pellagra that contributed to poor immune systems and failed pregnancies. A niacin (vitamin B3) and lysine deficiency, pellagra is a dangerous metabolic disease that can lead to failed pregnancy, loss of muscle control, constant diarrhea, skin lesions, insanity, osteoporosis, and, if unchecked, death (Gillman and Gillman 1951).[9] A few weeks of fresh vegetables and some meat can reduce some of the worst of the gastro-intestinal symptoms. Six months of proper diet cures a high proportion of sufferers.

A modest consumption of wild turkey meat would have reduced or eliminated the risk of contracting pellagra. Of all wild meats, turkey best ameliorated niacin deficiency. It is unknown when the long-lived cultural relationship between turkey and small foraging settlements first emerged. We do know that the relationship endured for thousands of years, leading to the domestication of turkeys. In later Chacoan times and successive Puebloan society, reliance on turkeys lasted right into the late 1900s AD.

Cultural Trajectory in the Middle Archaic Period

A distinct pattern began to emerge about 4500 BC in the Middle Archaic of the San Juan Basin. Broad-spectrum foraging/storing groups began to dominate the archaeological record by roughly 3500 BC, according to site-type records filed in Santa Fe's ARMS database.

Hunting populations and their camps, however, appear not to have grown in size or number during the Middle Archaic Period. Over time, hunting groups had been unable to fully break out of their seasons of want. Narrow, specialized food targets like big game was a risky long-term dietary strategy in an era and region noted for its climatic swings. Large hunting camps have been documented until roughly 2000 BC in the Middle Archaic, when the very last ancient-style bison hunters either disappeared, moved north, or adapted and became foragers.

Thereafter, plant processing and storage became the cornerstone of family security among the majority of the San Juan Basin's forager populations and opened the path to later gardening practices. Indeed, the San Juan Basin region was inching toward a culture of gardening and horticulture. The earliest San Juan Basin's tiny-cobbed corn plantings found in the Chaco Canyon region are dated to about 2000 BC.[10]

CHAPTER 3

THE ROBUST PLANT AND FORAGING SOCIETY OF THE MIDDLE ARCHAIC PERIOD

DURING THE MIDDLE ARCHAIC Period (3300–1500 BC), women's cultural and economic value most likely rose thanks to their increasing knowledge of plants and food preparation techniques. They were also the primary grinders of grass seeds, acorns, etc.—hard, exhausting work, and crucial to family security. Women were also likely the primary makers of basketry at this time. Sources of basket material came from beargrass (*nolina microcarpa*), sumac (*rhus trilobata*), and narrow leaf yucca (*yucca glauca*).

Children's workloads also enhanced food security among broad-spectrum foraging families in this era. By age five or six, children of both genders went out in early spring to forage grass seed. Their caloric returns of six or seven daily hours of collecting grass seeds or chenopodia leaves augmented the work of their elders. Plant foraging children could thereby begin to "pay" for themselves—gathering food surpluses that exceeded their own food consumption—and contribute to the volume of food in their family's storage pits. Thus, their value was much higher in a plant foraging society than in an ancient hunting band. Moreover, among the plant foraging children there were more potential mothers—a huge factor in the region's population growth.

Children's hard work also played a major role in shaping their growth patterns and later adult height (Fogel 2012).[1] Indeed, a regional reduction in child birth weight, stature, and longevity appears to have resulted as foraging became more widespread and children worked harder.

Seasonal Camps and the Genius of Female-Created Landscapes

Between 3000 and 2000 BC, hunting roles became secondary for San Juan Basin foragers. The increasing number of broad-spectrum foragers' winter camps drove the larger herds of deer and elk into the vast, snow-covered territory above the tree line. Hunkering down in late fall in a favored ecotonal location, families still worked to fill storage pits of piñón nuts, acorns, and late-season berries, as well as gathered small game meat to dry. A layer of chopped or broken pine boughs and other plant detritus made for a decent sleeping platform, partially protecting family members from the cold winter ground. Juniper and pine boughs provided a brushy roof overhead, fronted by a small stone hearth.

A few centuries later, the natural deterioration of those brushy roofs would help precondition the winter campsites that became small, three-season gardens of highly diverse plantings, a legacy bestowed by the first transient families that had returned to them seasonally, leaving their temporary camps richer for the generations that followed. Many more families began to live in old camps as their primary homestead, a dynamic evolutionary trend.

Regional Societies Form and Tensions Grow circa 2200 BC

If spring came quickly and gently, providing lush expanses of Indian ricegrass and chenopods, deaths and failed pregnancies could be minimized. If it came late, bringing unexpected snows and deep cold lingering into April, those who were ill or had high metabolic rates and low body fat were at most risk of death. As in earlier periods, that often meant infant children and elders.

Under the circumstances sketched above, it would have been essential to have support relationships with other nearby broad-spectrum foraging family lineages. As population grew, cooperating multifamily foraging groups could join together to make local labor pools capable of rapidly maximizing plant and seed harvesting. More on these "sharing networks" later.

Relationships between plant foraging families and hunter-foraging groups in the San Juan Basin were formed, too, and likely endured through the Middle Archaic Period (about 2200 BC) and even later. It is even likely that some forager-born females married second-tier hunters. Such men could hunt, enhancing the family's protein intake, make tools, and provide a margin of protection for the family with their atl-atls.

Archaeologists do not currently have enough evidence to nail down the precise trends nor make numerically explicit statements, but we know that relationships between hunters and foragers were also tense, mainly involving food competition. As population increased by the end of the Middle Archaic and the beginning of the Late Archaic, regional hunting pressure collided with larger, more efficient plant foraging teams. Adding to the tension was the fact that broad-spectrum foragers also hunted small game.

The consequences of increased food competition, as always, were multifaceted. By roughly 2500 BC, average regional temperatures had declined, and precipitation had increased—that slow process produced a more expansive plant groundcover and somewhat higher levels of plant diversity. With increased groundcover, more small animal species prospered, and both hunting families and broad-spectrum foraging families benefited (Roth 2018). But as population continued to rise in the San Juan Basin and its encircling high country, the benefits of somewhat more generous food resources eroded.

Centered around the tension between hunters and foragers was access—exclusive access—to more lush foraging or hunting territory. Added to this was the need for more ways to process and create spoil-proof food (Hanselka 2018). Successful food storage was key to a family's security in the face of near-starvation in late winter/early spring.

Most vegetable storage pits are assumed to have been grass lined. Grass linings needed to be clean and dry—even one wilted leaf or bruised chokecherry could have tainted an entire winter season's food storage. Storage pits needed to be dug in fairly dry soil. No wet, moldy food could be stored; the lid had to be airtight. A similar practice of filling root cellars and backyard dugout pits continued in use in the western Pennsylvania "Coal Patch" towns right into the early 1900s (Stuart 2019).

Forager camps with ample stored plant foods presented rich targets for unsuccessful hunting groups whose bowmen could easily cower a single foraging family. This caused many foragers to move to more secure hillside "winter" camps, which were hard to approach unseen. The best locations for such camps were on south-facing vegetated hillsides at about fifty-five hundred to sixty-five hundred feet in elevation. Such a location maximized exposure to fall-winter sunlight, benefited from the hillside's warm updraft effect, and offered access to the rich, diverse piñón-juniper ecotone. Well-sited camps like these grew in size and numbers throughout most of the Middle to Late Archaic, roughly 2500–1500 BC.

Challenges Faced in the Late Archaic

By the Late Archaic Period, regional foraging society had gravitated to dozens of subtle but distinct plant resource targets: mixed grasslands, sagebrush flats, cactus groves, piñón groves, juniper or pine zones, and the richly diverse ecotones where plant and animal resources collided. The success of broad-spectrum foraging relied on biotic diversity, ample acreage to exploit, as well as a wide variety of food-procuring strategies, which we can identify from the diverse toolkits found from camp to camp.

Finding the most food-productive seasonal campsites and maintaining "rights" to those campsites allowed for stability among foraging families, particularly as the task of provisioning for winter became more difficult as regional population continued to grow.

> **WHAT IS A TOOLKIT?**
>
> An inventory of all portable artifacts found at an archaeological site is generally referred to as an *assemblage*. Different cultural groups have different assemblage profiles. In the Southwest, the dominance of flaked stone artifacts often points to hunter-forager groups, while a preponderance of ground and abrasion-shaped stone artifacts points toward broad-spectrum foragers. In an exposed site with artifacts scattered across eroding hillsides or dune blowout an experienced pair of eyes can assess most Archaic Period sites as hunter-forager or broad-spectrum forager in just a few minutes. The total assemblage of these artifacts found in a given site is the "toolkit."

As competition for food security continued to increase, heavy foraging decreased plant diversity in some of the drier districts. In addition, an arid season or several years of spotty, irregular summer precipitation could no longer be remedied by simply moving on. Each generation faced fewer landscape foraging opportunities, further placing a premium on controlling access to food-rich plant ecotones. Intensifying strategies designed to produce more food also became more important: careful weeding of inedible invasive plants from seasonally foraged hillsides; light weeding to increase the growth of wild amaranth and *chenopodia*; and enhanced efficiencies in the processing of acorns, juniper berries, sunflower seeds, roots, and berries.

Clumps of amaranth signaled moister soil beneath the tangled groundcovers common to the region. Amaranth seed wasn't just a vitamin and protein-rich food—it was one of nature's most reliable tells of moist soil pockets and disturbed soils, tempting families to settle there. As newcomers came, original inhabitants claimed those mid-elevation family camps as "home camps." And it was likely in those moist locales where wild amaranth grew that women's small, prototype gardens were first created in the Late Archaic.

TIERRA DEL FUEGO ISLAND, SOUTH AMERICA, 1885-1920: A STUDY IN SELF-EXTINCTION

A comparison to the likely consequences of cultural collisions of land use in the San Juan Basin is revealed by a twice-documented case study of competitive population dynamics. First published was Lucas Bridges's book, *Uttermost Part of the Earth* (Bridges 1948), followed by my 1972 doctoral dissertation, "Band Structure and Ecological Variability: The Ona and Yahgan of Tierra del Fuego."

The Ona hunters' ancestors likely arrived on their subarctic island about nine thousand years ago. Ona society consisted of about a dozen bow-hunting bands. The size of each band was carefully regulated by each group's powerful headman and shaman (holy man). The headmen were first to emerge from their winter camps to assess the size and location of *guanaco* herds. Guanaco are one of two wild South American camelids; the other is the *vicuña*. The headmen's assessment led to planned hunts with elegant longbows for guanaco meat and hides, and binge-eating to regain body weight lost to winter's stingy meals.

Suddenly, in the late 1880s, on the southern tip of Earth, the aggressively growing country of Argentina presented the ancient Ona with a desperate situation induced by the import of huge flocks of European sheep, which grazed the vast grasslands of chest-tall, native wild grasses down to the stubs. The ecological destruction of these grasslands rapidly reduced guanaco populations, a grave threat to the food supply upon which upland Ona bands relied to exist.

In response, Ona hunters used their bows to kill invading sheep, which they did not eat. Enraged sheep station owners responded by killing Ona with repeater rifles. To encourage the killing of Ona, these sheepmen posted cash bounties on pairs of Ona ears. Worse yet, new European diseases to which the Ona had no immunity began to sweep through those bands living nearest the sheep stations.

With a food shortage affecting their populations, Ona headmen responded by calling off the fall rituals that had annually initiated young Ona boys into manhood—those rituals had always included the right to seek a female sexual partner. In short, the Ona headmen had put a "stop order" on population growth. The ecological triggers of declining guanaco had demanded a declining population. This would not prove a successful strategy. Within about five years, Ona bands had begun to turn on one another, stealing each other's women and land.

When the Ona-on-Ona raids ended by about 1900, only a few hundred were left. By 1920 the few remaining Ona men worked on the sheep stations. Ona women were few in numbers; there was no stable role for them in the sheepmen's world. The combination of a rigidly organized society focused on a primary food—guanaco—and unleashed rage upon each other had undone nine thousand years of stability in just three decades.

Opening the San Juan Basin to Gardening

By about 1300 BC, the San Juan Basin world had shifted to a complex combination of foraging, small-game hunting, and small-scale gardening/horticulture. Old-style atl-atl hunters had been partially eclipsed as the big game they had hunted for millennia had declined, replaced more and more by foraging and gardening practices. Broad-spectrum foragers settled down and began to dabble with early horticulture.[2] Garden plots supported more people on smaller swaths of land, alleviating potential "all or nothing" demographic warfare, such as the one that ended the Ona world on Tierra del Fuego.

Comparing the well-documented ethnographic example from faraway Tierra del Fuego is helpful to understand the dynamics of population density sensitive behaviors. As discussed earlier, the long-lived hunting society on Tierra del Fuego illustrates the reality of how a society's cultural, population, sex, and density rules can either protect their well-being and food security in an era of intense competition, or destroy their society by turning on one another. Misreading the portends to an ecological disaster can be, and has often been, a cruel factor in the disappearance of many ancient and classic societies. Ignoring the key role played by constancy in female value systems is another source of cultural instability.

In New Mexico's San Juan Basin, female horticulturalists first created small, diverse gardens that handily out-produced the natural caloric yield of a wild quarter-acre of mixed grass cover by about 1500 BC. That scenario likely helped to reduce tensions between hunters and gardener-foragers during the Late Archaic Period, but as population rose over the next few centuries, violent conflict once again flared up on the northwestern frontier of the San Juan Basin in about 600 BC (Whittaker 2012).

The earliest horticultural districts identified in the San Juan Basin first expanded on its western margins where water tables were replenished, most years, by rain and snow-melt runoff from the Chuska, Carrizo, and San Juan Mountains and an arc of creeks, small rivers, and springs that flowed on the southern flanks of the Basin's limits along the two Puerco Rivers, west and east, and along the Rio San Jose. Those fairly reliable streams provided water to the region now defined as the Acoma Pueblo and Laguna Pueblo's sovereign lands. These are the districts I identify as "most valuable real estate" throughout this volume. In other words, part-time farmers and horticulturists in what are now the Acoma and Laguna districts were likely more independent, healthier, and a bit taller than people who settled in the districts of crop-risky, spotty rainfall and even riskier dry-farming.

Geography alone did not determine the success or failure of early small-scale horticultural experiments. The frequency, quantity, and timing of precipitation played a massive role in annual plant food harvests, as did annual temperature swings and distinctive soil types. Skill and patience also played a singular role in bringing processed wild plants to the fall storage pits. Thus, the detailed techniques of food storage still mattered greatly—one moldy ear of small-cobbed corn or a single rotting berry could ruin the contents of an entire storage pit filled with several hundred pounds of dry food, leaving an entire family at risk of starvation during late winter and early spring.

An Era of Technology and Experimentation

Unlike Tierra del Fuego, the mode of food-getting behaviors changed notably during the Late Archaic Period in the San Juan Basin (Kearns 2018), when the number of known archaeological sites increased rapidly. Technological experiments flourished: grinding stones grew in surface size, small bone tools proliferated, made from rabbit bones, migratory birds, wild turkey, deer, and pronghorn or desert sheep. Between 2000 BC and 500 BC, archaeological data indicate a pronounced trend in steady technological improvements in everything from tiny bone awls to seed grinding stones to larger yucca-fiber carrying baskets. Human hair continued to be woven into rabbit trapping nets, even as yucca fiber use expanded. Some wild cotton fiber was also processed into cordage and cloth.

The era was one of experimentation and technological adaptations. Daily toolkits found in the northern, higher elevations of the San Juan Basin differed from those found on the east face of the Chuska and Lukachukai Mountains on the western boundaries. Many regional toolkits included tools made of glassy obsidian stone found in the huge Jemez caldera.[3] This ecologically rich area also nurtured both an ancient herd of genetically distinct elk and large forests rich in rabbits, squirrels, deer, wild turkey, blue grouse, and big horn sheep.

In contrast, the sandy loams of the south-central San Juan Basin were hotter and drier than in the north. Groundcover was thinner, game scarcer, and horticulture riskier. Thus, one finds detectable stylistic differences in gardening techniques, bone and stone tools, woven textiles, and daily lifeways between these districts. On the basin's northern borders, larger game animals like deer, elk, or mountain goat provided more bone tool material than in the drier central basin. There, jackrabbit, cottontail rabbits, and occasional deer or antelope bones prevailed. Bison kills in the basin were very rare by this period.

Declining food sources stimulated women's experiments to produce food more efficiently. Regionally distinctive toolkits reflected multiple food-getting pursuits in differing altitude zones and in ecologically different locales—for example, piñón-juniper zone camp tools versus ponderosa and acorn high-altitude zone toolkit and styles of grinding stone. Seasonal resources at 5,200 feet in elevation and at 7,500-8,000 feet were quite different, as were temperatures, precipitation, lengths of seasons, targeted food sources, and tools.

Sharing Networks: Organized Self-Help Institutions Blossom

Differing toolkits and larger, more numerous camps defined the growing intensity of land usage during the late Middle Archaic Period. Among foragers, the increase of resources often depended upon long-established regional family alliances called sharing networks.[4] Sharing networks fostered flexibility, diminished hostility and competition, and enhanced a small family district's access to other ecozones. Such networks connected families over long distances, moved emergency food to famine spots, and carried information about rain, soil, or even grasshopper swarms. Every child born extended this crucial, long-distance information system until its mobile linearity formed a lattice work of human agents intertwined in a shared endeavor. Cooperation across swaths of landscape enhanced social order—a calming and entropy-suppressing cultural practice. Interfamily trade relationships also gave isolated families an important sense of belonging.

Sharing networks would have suited hunters as well as foragers. In the Southwest's Great Basin, raiding, wife stealing, and withholdings of male initiation were all tools that could have been used to diminish hunter population density. In the western forested margins of the San Juan Basin, several sources document an increase in penetrating wounds in the late 400s to 100s BC as the Archaic Period faded away, rendering sharing networks even more important.

Family-to-family trade relationships worked best when trade partners gardened on different soils, lived in varying elevations, and grew slightly different corn and squash varieties—food diversity was a powerful buffer against starvation. Multiple trade partners were better than just one, though making promises to too many trade partners was risky, as dry years limited the volumes of food sharing.

The Pioneering Gardening Women

The women of the Middle-to-Late Archaic were the ones who ground much of the early crops of hard flint corn and carried water to the family garden. Both of these work roles were economically and demographically valuable to group food security among extended foraging-gardening families.

As in more ancient times, many young women of this era likely lost at least one child during their brief reproductive years. In the 200s–500s BC, the mean age of female death was about twenty-one or twenty-two years (D. Martin 2000). Since their first menstruation was around age eighteen, many young women had only a few chances to bear a healthy child.[5]

Over the long, complex period of 3500–500 BC, hunters, foragers, and emerging gardeners protected their distinct cultural worlds. Differentiations in preferred foods, daily routines, and cultural visions of normalcy all contributed to the rise of distinct subcultures in the San Juan Basin. Linguistic anthropologists do not know what mix of specific languages were spoken in the San Juan Basin in the centuries leading to AD 1, but one can assume that several languages arrived as newcomers drifted into the basin, carrying new varieties of corn, beans, and squash.

Some hunters and foragers intermarried and cooperated in food sharing networks. This undoubtedly pushed some broad-spectrum foragers out of their favorite upland hunting districts.[6] The goal for hunters was to reinstate their vision of "safe" human-to-land ratios amid rising foraging populations.

On the more positive side, more robust vegetal storage, plant harvesting, and food storage in women's emerging proto-gardens of the Late Archaic Period (1500–1000 BC) softened regional famine risks and therefore reduced the impulse to indulge in open conflict. Some evidence of this is available from the Black Mesa district of Arizona, where violence appears to have declined over several centuries during the Late Archaic Period.

Changing Patterns of Women's Land Use in the Late Archaic Period

The upper piñón-juniper ecotone ranged from 6,000 to 7,400 feet in elevation. This ecozone was a diverse food territory, and generated competition for seasonal access. In the lower and warmer altitudes (5,000 to 5,800 feet) of the same zone, desert plants mixed with juniper, cottonwood, pine, and piñón. Ancient families had long moved systematically from through high to low country in the course of their regular seasonal migrations, but by 1300 BC, an era of rapidly rising regional population, access to all the traditional foraging zones had become more difficult—families with larger base camps began to dominate foraging in the most highly productive areas.

Thus, the reality of rising population in the San Juan Basin during the Late Archaic required even harder and longer hours of foraging. Everyone aged six to thirty worked, with the exception of grandparents. Internationally acclaimed anthropologist Hillard Kaplan and his colleagues' research in South America indicates that, in general, plant foraging groups grew in number partly because of the need to support multiple generations in each camp. As some adults aged into their thirties and forties, a number among them would have suffered enough leg pounding, back stressing, foot bruising labor to limit their speed, strength, and leaf/seed stripping work capacity. Even their teeth would have been ground down by decades of chewing sweet, carbohydrate-rich plant stems and the stone grit that made its way into ground, plant-seed meals.

Multigenerational forage groups needed to protect those aged members because they were crucial knowledge bearers. Aging members of broad-spectrum foragers dictated slower walkabouts, shorter distances between seasonal camps, and the cultural trait of sharing. In contrast, children six to ten years in age were useful, especially as their skills improved, but did not reach their full capacity for carrying heavy loads until their early-to-mid teens. By age ten, the boys would already have learned how to throw their curved wooden rabbit sticks, as did most of the women.[7] The atl-atl still ruled for hunting and defense of foraging territory and stored food.

Jemez Cave

As the San Juan Basin's population rose, so did the size of family groups and their camps. Compared to the 2000s BC, when the basin's oldest documented corn was sown, the region was bursting at the seams by 700–500 BC. Broad-spectrum foragers needed more space in which to forage, as well as access to the most food-productive ecotones, like the upper piñón-juniper belt. One of the most important archaeological sites of this era was Jemez Cave, where very early squash and small-cobbed corn had been grown seasonally. Jemez Cave's corn samples were dated to 440–410 BC (Vierra and Ford 2006).

Jemez Cave is important, as it has demonstrated evolutionary trends in a region on the forested eastern margins of the San Juan Basin. Those trends included early planting experiments with both squash (small-sized) and early corn (very small-cobbed). Dr. Richard Ford assessed that the early squash and corn were sowed, but only minimally tended, suggesting that the seeds were planted, and then the cave was abandoned during the summer while its occupants continued on to their traditional plant/small-animal foraging rounds. They would then return to the cave in mid-fall with the hope of corn waiting to be harvested (Vierra 2006).

Jemez Cave strongly shows the ambivalence among broad-spectrum foragers about investing time and heavy labor in early corn horticulture, as opposed to setting out on foot each spring and summer to forage. That ambivalence makes sense since the available corn varieties were very small-cobbed and hard to grind into meal. The higher elevation of Jemez Cave (at 7,062 feet above sea level) would also have limited the corn's growing season. Despite this purposeful planting, it must have been more productive to forage widely at this time (400 BC), even when a snug upland cave remained in a group's possession for plant storage.

Thus, the Late Archaic Period presented a quandary to many ancient families: Which foodstuff to pursue, and where to find it? Hunting produced the most calories and protein per unit of labor, but competition for hunting rights had become turbulent and large game were scarce. In contrast, foraging was more reliable but returned fewer calories and protein per work hour. Slowly and steadily, women's gardens breached the gap by selecting premium wild plants as core foods and domesticating the best and largest corn, squash, and beans.

Rapid Cultural Changes

The period from 1000–500 BC was the era during which the thermodynamics of the San Juan Basin changed irrevocably. It first came in the forms of tense food competition, increased reliance on sharing networks, and the use of small family gardens to harvest more grass seeds, including the fortuitous semi-domesticated plant species like amaranth, hybrid grasses, and sunflowers. A wave of new varieties of corn and squash made their way to the San Juan Basin from southern Arizona and northern Mexico in the period between 2000–500 BC. Some of those seeds were a small-cobbed variety of *zea maize*, which we know as "corn." With its arrival came several varieties of squash. With these arrivals came the potential to enhance food outputs, engineer new landscapes, and mimic natural food-producing "plantscapes" to feed a steadily growing population.

New human-altered landscapes and food production efficiencies depended, in part, on early corn's slowly increasing cob size. Additionally, sowing corn in plots that had previously grown squash and beans enhanced the nitrogen in the soil and offered higher dietary protein content when all three were eaten together. Daily life became more complex as survival strategies multiplied. In ecological terms, the dynamic-static balance scales noted in a preceding chapter had begun to shift from a heavy reliance on hunting with occasional plant foraging to foraging alongside the creation of

Early Zea maiz cob (2.5 inches). Drawing by Baker Morrow.

proto-gardens. Between 1000 and 500 BC, the world of the San Juan Basin had grown more complex as small, regional subcultures tested experimental lifeways. These included multiple food strategies, life in different climatological life zones, rapidly changing landscapes, profligate use of wood, and an ever-hungry, growing human population. Planting corn was about to change the ancient San Juan Basin world in ways that would shape society for the next twenty-five hundred years.

PART II
CORN & WOMEN'S GARDENS

CHAPTER 4

THE GENIUS AND INNOVATION OF ANCIENT WOMEN'S POCKET GARDENS

It is mid-March in the year 336 BC in the San Juan Basin. The icy spring winds howl. It will be cold on the mountain, a mother tells her young children. Their grandmother has made gruel with the last of the fall seeds harvested from her small garden's two large amaranth bushes. Grandmother's garden has saved them from starvation so far this winter. Her daughter, granddaughter, and two grandsons bundle up in layers of thinly woven cotton and tree fiber rags. Yucca fiber sandals receive a warming layer of mashed leaves, and their feet are swathed in loosely woven wild cotton and scrap leather. They eat the last of their food and pray that it does not start snowing. Their faces are bronzed by the harsh, high-altitude sun.

They exit their camp, a humble brush-roofed dugout dwelling tucked under trees at 6,800 feet in elevation on the east-facing Chuska Mountains. They climb uphill till they reach a clearing that overlooks the Chuska Valley. There, they meet several other adults and children—the other members of their shared food network. The thirteen adults and five children all look gaunt, as their body fat has been reduced by tiny winter meals. They walk together and arrive at the face of another steep hill. Nearly thirty acres of dense, gray-green Indian ricegrasses carpet the hillside below the equally dense tree line.

By three in the afternoon, the hill has been stripped of grass seed. Nearly seven hundred pounds of seed are put in twenty-five baskets. The men's baskets hold about forty pounds and the women's around twenty-five. The group's ranking female remembers that in her childhood, this same hillside produced about eleven hundred pounds of grass seed—almost double today's bounty.

The women have brought their throwing sticks and kill several rabbits. As the basket carriers descend, three more young men arrive with dogs and atl-atls to protect the portion of hillside yet to be harvested.

DIET AND PREGNANCY IN UPLAND VERSUS LOWLAND FORAGERS

On May 16, 2024, an article with the heading "Making Babies May Take 10 Times More Energy than We Thought" was published by Lauren Leffer in the online journal *Popular Science*. The new research and conclusions outlined in the article help to explain important distinctions in pregnancy and population dynamics between the upland and lowland foragers in the San Juan Basin during the Archaic Era.

For upland foragers, ricegrass not only supported their daily labor but also played a seminal role in the spacing of their pregnancies. The high-protein diet offered by the grass seed actually modulated births by generating periods of roughly *three years* between pregnancies. That ample time allowed mothers' bodies to be fully restored before the next pregnancy. In short, high-protein seed diets were an ancient, probably unnoticed actor in population dynamics—fewer and more spaced-out pregnancies diminished group food needs and increased efficiency. More details about this are published in Bryan Sykes's book, *The Seven Daughter of Eve* (2001, 267).

By comparison, in the lower elevations of the San Juan Basin where high-protein foods were scarcer and diets consisted mainly of carbohydrates, foragers' pregnancies occurred in shorter time periods. More frequent births stressed their food supplies, tying them to a more sedentary and stressed lifeway than the upland foragers.

Thus, we have at least two populations that were, in energetic terms, living in totally different biospheres separated by only six to eight hundred feet in elevation! Upland foragers were metabolically efficient. Lowland foragers were demographically more powerful but less food secure.

The Shift to Gardening

Throughout the last half of the Late Archaic Period (1000–500 BC), forager groups' annual movements, harvest selections, and minor landscape alterations to expanses of grassy landscapes had significantly enhanced the region's diversity of edible plants. Annual weeding generated a slowly declining profile of undesirable, invasive, and inedible plant species in these seasonal campsites. In shaded, but open, upland plots where wild amaranth thrived, some broad-spectrum foragers began to experiment with small-cobbed corn called *chapalote* at this time, rather like that sown at Jemez Cave.

These human-curated plants became small proto-gardens where soil alkalinity, mineral nutrients, and fresh water sources were present. Grass baskets smeared with pine pitch were used to carry water to camp. All of these conditions and practices combined into a very complex, learned suite of plant ecology and the assets of a desirable campsite. The knowledge required increased human work inputs and resulted in reliable plant husbandry, seed storing techniques, and gradually declining seasonal movements.

These grassy, seed-rich patches were being shaped during the Late Archaic Period into experimental pocket gardens, curated by women. Many of the gardens' earthen floors were dug low enough and modified to capture runoff moisture; brushy windbreaks were built to protect young plants from late spring's cold, east-to-west winds. These early gardens ranged in size from an average modern bedroom to a three-car garage. Larger gardens suggest several generations of development and growth. Where needles or leaves fell from nearby juniper trees and bushes and decayed, both wild and edible plants thrived.

The emergence of women as authors of the first gardens in the San Juan Basin was not unique. It was part of a worldwide hunter-forager pattern where large animals had been overhunted. Ancient women seemed to see and become part of a plant and small-animal ecosystem that most ancient men did not enter into. Women's curated ecosystems focused on quality: nurturing diverse plants and harvesting small animals as a way to feed and nourish families more wholly and reliably. In contrast, the males tended to pursue larger quantities of food in the form of big-game hunting, even when it was futile.

Pre-Chacoan pocket garden: corn with beans. Drawing by Baker Morrow.

With women at the forefront, the Late Archaic Period's forager groups steadily "mapped" themselves onto the local landscapes, heeding seasonal precipitation patterns, as unpredictable as they were, as well as soil quality, edible plant diversity, elevation, crop success, and the needed volume of storage capacity. Assessing seasonal outcomes of successful or disappointing crop hauls went on for roughly fifteen hundred years. The slow, choppy experimental process informed small but continual advances in seasonal gardening. Gardeners came to understand that in a "perfect" year, abundant monsoon rains came in August/September, followed by several late fall rains, then winter snows. Snowmelt from mountain snow fields moistened lower-elevation campsite soils into the months of April to June. In many middle to upper elevations, these snow packs fed seasonal rivulets, along which casual gardens could be established.

Over time, women foragers and their proto-gardening sisters learned they had to grow and store enough vegetal foods to weather more than one season of unpredictable precipitation. Nothing was guaranteed, but certain cultural behaviors did buffer risk: store more, eat less, grow garden near water, forage, and continue to participate in emerging sharing networks that connected gardeners living in differing ecozones. These behaviors were crucial to the transformation of the San Juan Basin's once-isolated local populations into family groups that engaged on a regional scale.

Gardening Rules, Risks, and Complexities

The rules of movement pertaining to each hunting-foraging or seasonal gardening group changed by the day and, at times, by the hour. Moreover, different sets of rules applied to foraging-gardening groups who became sedentary for several months each year. The more successful gardeners with access to premium ecozones (clayey loam soils), high natural plant diversity, and occasional small game tended to extend their summer and early fall seasons in one place, generating a trend to increasing sedentism, which would eventually become a key change for foraging groups' lifestyle.

Early inhabitants of the San Juan Basin had to assess caloric costs and returns on work investments daily. Planting and weeding was tedious work. Carrying water to a forty-foot square patch of corn in a bone-dry July cost many human work calories: the walk to a water source; the return to a plant or garden patch; the weight of the water carried; and some percentage of the initial work calories to make and waterproof a basket with pine pitch. The fetching of water cost time, planning, tweaks to baskets, and cost many human dietary calories—"work," as we define it. Thus, areas closest to *ojos* (springs or seeps), *ciénagas* (swamps or bogs), or near semipermanent moist streambeds and rivulets became highly sought-after territory.

Experimenting with corn growing was a time-consuming endeavor. Corncobs dated to 800–500 BC were small and hard to grind. Worse yet, reliably identifying the specific varieties of seed corn was an alchemist's art, and keeping track of the more successful corn races was difficult.[1]

Corn plots on land with ground water in the form of seeps, springs, or tiny seasonal streams reduced the risks, because successes depended much less on the vagaries of summer rainfall. What few individuals who lived in the San Juan Basin between 1200–500 BC apprehended was the new complexity of their regional society, which had split into three main food-producing trajectories in just a few centuries. Those trajectories included intensive plant foraging, traditional large-game hunting, and two categories of gardening: wetland and dryland.

Finally, the powerful, ancient rules of female "primacy of land use" left newcomers to the San Juan Basin to scrounge and fit in as best they could. Even today many archaeologists have not fully apprehended that ancient cultural adaptations surrounding land and foraging were tied to this primacy of land use. Women's lineages formed by prime land rights likely defined communities more formally than male hunters' designations of premium hunting territories. After all, game animals moved frequently, but women's gardens did not.

Hunting, Foraging, and Gardening Lifeways Develop

Recall that increasing complexity is a core attribute of a power phase. In the Late Archaic Period between 1500 and 500 BC, innovations abounded and lifeways began to change more rapidly in the San Juan Basin. By roughly 2000 BC, the three coeval lifeways were in competition for resource-rich locales. The Hilltop Buffalo kill site is a clear echo of very ancient hunting patterns, while the small-cobbed corn grown on the perimeter of Chaco Canyon in 2000 BC was a harbinger of foraging life that had developed new food strategies (Salfisburg, Cordero, and DelloRusso 2014).

Hunters produced the most food calories and protein annually, especially considering their return on work calories. Timothy Kearns provides data on the size of hunted game in the San Juan Basin throughout the course of the Archaic Period (Kearns 2018). Archaic hunters took small game from lower elevation grasslands. Those hunting the higher elevations ranging from sixty-two hundred to seventy-seven hundred feet were typically taking large game like elk and mule deer. By 1000 BC in the Late Archaic Period, large game animals were significantly declining in numbers.

Broad-spectrum foragers worked far more annual hours than the hunters, but rising regional population reduced per capita wild forage returns. One foraging group was not exactly like another. Among the San Juan Basin's ancient populations, specific food-getting strategies varied. Thus, rich pockets of both plant and animal diversity were changing in many foraging districts. By the Late Archaic Period, some family groups avoided forage-rights conflicts and began to plant gardens of small-cobbed corn, beans, or squash. These small upland gardens were weeded and tended till cornstalks and small squashes and gourds were established, then left to mature until a group returned to their garden camp in late summer. This is similar to the pattern found at Jemez Cave a hundred miles to the east of the San Juan Basin.

The garden-focused families worked harder annually than did the foragers, but only those gardens on watered ground—wetland gardening—produced consistent food returns. In contrast, dryland gardeners in open, low-lying valleys still had to forage most years. Imagine an upland cove of thirty acres filled with wild Indian ricegrass, amaranth, and *chenopodia* plants rimmed by forests of large junipers and piñón trees as roughly equivalent to Times Square in Manhattan. By roughly 500 BC, everybody from everywhere wanted to be there, all at once.

The Cultural Evolutionary Trajectory Toward Horticulture

In spite of the slow, irregular start, limited by hard-kernelled, small-cobbed corn, and unpredictable precipitation, core lifeways among family foraging groups in the San Juan Basin began to drift even further apart in the Late Archaic Period. Some decided to intensify their foraging. Others mixed seasonal gardening with foraging, and less fortunate newcomers farmed along the bases of dry hillsides.

Each of these nuanced foraging-gardening patterns were slightly different with respect to crop yields and a landscape's wild food inventories. Human labor required food energy, vitamins, and minerals. As population rose during the Late Archaic, the scramble for food and access to foraging rights again contributed to social tensions and may have led to notable changes in weaponry, family shelters, and patterns of seasonal movement.

By about 800–500 BC, both broad-spectrum foragers and horticulture had burgeoned. The seasonal camps of broad-spectrum foragers had grown in numbers and complexity (Kearns 2018). Architectural complexity came in the form of larger brush or branch shelters, more numerous hearth areas, and more refined storage pits. The largest of these archaeologically documented fall camps could have accommodated more than a dozen small family groups. In more poorly vegetated districts, a broad-spectrum forager's spring-summer camp cluster was smaller. Archaeologists know that by counting the number of family hearths.

By the end of the Late Archaic Period (ca. 500 BC), the archaeological traces of traditional hunter band camps were scarce. In their wake came more experimental gardeners, usually described as emerging "horticulturalists," and became sedentary.

As the basin's populations continued to grow, its residents had to work harder, so consumed more food calories per capita than in earlier eras. From flotation analysis of seed and plant remains, many very early gardens included traces of small-cobbed flint corn, amaranth, squashes, gourds, sunflower seeds, and *chenopodia* species like lambsquarters or goosefoot. Beans are rarely identified in these early gardens.

In the relatively uncrowded mid-elevations of piñón and juniper ecotone of about 1000 BC, pine nuts, amaranth and wild grass seeds, juniper berries, rabbits, and the squirrels that relied on them all provided a sufficient diet in most years.

Female horticulturalists' mid-elevation garden camps likely drove some large game into higher elevations. In response, hunters in lower elevations may have begun to move higher. Multiple competing lifeways complicated daily life. Among these multiple subcultures, female proto-gardeners and keepers of upland grass seed foraging camps would most heavily shape the future trajectory of daily life for at least a millennium in the San Juan Basin.

The early female gardeners and their female successors suffered many climatological risks over the centuries. Their history is still not fully documented. Yet they reshaped small bits of their landscapes to consistently increase food productivity in small spaces. Think of it—each successful female gardener made it possible for a few more humans to survive on a fixed acreage landscape, thereby reducing conflict and protecting precious grass seed foraging patches.

TABLE 3. Estimated Rough Dietary Calories Consumed in Gardening

Activity	Approx. Calories per hour per 120 lb. man	Approx. Calories per hour per 105 lb. woman
Light gardening	200–225	190–210
Plant foraging	160–175	145–160
Processing big game	250–325	300–375
Hunting rabbits, etc.	175–225	150–200

Transition to the Terminal Archaic, ca 500 BC–450 AD

Over time, land ownership became even harder for a newcomer family to achieve, especially a garden plot near water. Small, isolated family groups of six to ten people would have been relatively easy to evict from their up-country summer/fall gardens; even a few days' eviction from tending a several-week-old planted corn patch could ruin that season's entire corn crop. A small family group also presented an easy target to fall or winter raiders who could target family storage pits. In the course of tinkering with horticultural style gardening, it became obvious that bigger family groups were better, both in the realm of available foraging and gardening, and in the security of their numbers. Several camp dogs were also worth feeding.

The Terminal Archaic was a period of frantic tinkering with small, intentionally planted gardens, and lasted about a millennium, but was successful enough to create the underpinnings of a modest, female, intense work–based power phase. That success was due in large part to the growing number of female gardeners. During a human-induced power phase, a flood of work calories are burned. Stoop field labor related to gardening would have included selective weeding, breaking the soil with fire-hardened digging sticks, punching many holes for the deposit of corn kernels, gently tamping the soil, watering it, then moving about three feet to deposit more kernels, and so on, until twenty to sixty holes had a kernel or two in each.

In a dry spring, pine-pitch-basket carriers likely poured water in each hole. None of this garden work guaranteed a healthy cornstalk to sprout from each hole. For that reason, other members of a proto-farming family group would have been nearby, harvesting and processing Indian ricegrass seeds as they had done for millennia. Grass seed collecting was a tough habit to kick, as it was essential for its dietary protein and iron.

In landscapes where natural vegetative tangles included both Indian ricegrass varieties and bordering clusters of *chamisa* (rabbit brush), other family groups could deploy their rabbit sticks, while aged folk and children stripped ricegrass seeds and chenopod leaves into woven yucca baskets. Chamisa played a role in proto-gardening, as it thrives in disturbed areas and is a plant that landscape scholars associate with gardening.

In short, there was no magical day, year, or decade during which the

Rabbitbrush sketch by Esther Burton.

new world of experimental horticulture replaced traditional land use and broad-spectrum foraging. It was a process full of fits and starts, failure and disappointment, and yet some notable triumphs as the techniques of planting in local soils piled up. Much new knowledge had been earned by what we now identify as "sweat equity." Millions upon millions of dropseed grass calories, along with women's small and carefully tended pocket gardens, had shaped horticulture as a dominant lifeway in the San Juan Basin by about 500 BC.

Women's Sophisticated Horticulture in the Late Archaic

The San Juan Basin's evolving world was rather like a ninth-grade biology experiment focused on the slow growth of microbes in a petri dish, followed by a sudden microbial explosion. In the San Juan Basin, it had taken centuries for the cultural microbes of growth, innovation, and knowledge to be analyzed by its human practitioners and successfully deployed. The archaeological record makes it clear that those who added garden cultigens to their foraged diets lowered their risk of famine.

This period of transition to gardening is difficult to analyze archaeologically, as many of these campsites are no longer intact. But gardening gave the descendants of broad-spectrum foragers several distinct avenues by which to pursue nutritional success: foraging, planting, or both. Their plantings added calories, sugar (corn), protein (beans), vitamins (squash and sunflowers), and water/storage vessels (squash or gourds). Eventually, the traditional upland spring plant foraging rounds became secondary due to declining grass acreage. Where surface water was reliable, early corn horticulture was successful. Early corn farm districts were founded in localities like Skunk Springs and Peach Springs in the Chuska Valley.

By about 500 BC horticulture became the most important regional source of calories and storable foods. Ancient foraging practices endured, but yields continued to decline. But between roughly 800–400 BC, a period of somewhat more abundant precipitation appears to have coincided with larger families, larger gardens, and more elaborate upland family shelters. This pattern hints at ownership claims to long-held garden plots. By the end of the Late Archaic, small-scale women gardeners appear to have dominated because of their families' superior diets.

Archaeological clues point to a dramatic surge in population during the Late Archaic Period (500–300 BC). New corn varieties became available both through trade and local cross-breeding experiments. More imported varieties of cultivated beans were available. Beans increased dietary protein and also recharged nitrogen levels in soils where one season's corn plantings had consumed the nitrogen needed to support next year's corn crop. The remedy was to shift corn rows by six to eight feet every other year, as well as to continue filling food storage pits with foraged foods. These needs would have also favored lineage-based claims to early spring foraging territory. A clear lineage of the names of female, garden-owning forebears may have become essential to retaining and owning garden lands.[2]

Power Phases

The beginning of what I label as a power phase is often plainly announced in human societies: diplomatic trash talk between the US and Russia during the Cold War days qualifies as an example. Our modern nation's penchant for breathless "breaking news" energizes emotions and daily behaviors, much as did the Confederate South's constant newspaper denigrations of "greasy Northern factory workers unfit to vote, who desired to invade the South" during the run-up to the Civil War. In short, human power phases define the rules of cultural combat and fit perfectly with the title of Clint Eastwood's Old West shoot-'em-up film *The Good, the Bad, and the Ugly*.

A modern ice hockey game in which a player's time on the ice consists of a "shift" of about 40–120 *seconds* is a power phase. The physical exertion of that hockey player is measured in *hundreds* of calories expended per shift, not our standard calories per work hour. Highly tuned athletes can burn as many calories a minute as most of us spend in half an hour of ordinary work. The evolutionary catch, of course, is that though human power phases promote great change, they have a short half-life. This can be seen in the rise, notable complexity, stagnation, and eventual fall of classic empires like Rome and Byzantium. The more efficient societies typically have longer lives than the more powerful. They are also less stratified, both socially and economically.

Ironically, the atomic/hydrogen bomb is New Mexico–stationed physicists' contribution to the ultimate human-created power phase . . . one so powerful that it renders its own matter and everything in its path into minute fragments of its former structure. Please note, history suggests that the cultural quest for ultimate power is typically short-lived and immensely destructive. While we may not be able to predict the future effects of the changes we make to our society, according to Christopher Langton, we are responsible for the consequences (Waldrop 1992).

As calculated for my book *Pueblo Peoples on the Pajarito Plateau: Archaeology and Efficiency* (Stuart 2010), a single traditional hunter/narrow-spectrum-gatherer of about 135 pounds (robust for his era) would expend about 672,500 food calories per year (or 1,843 calories per day) making a living. His high-protein diet probably gave him an adult height of about 5'5".

In contrast, an individual broad-spectrum forager male of the same period with a body weight of 125 pounds (less rich breast milk and protein in childhood) would have spent about 746,750 calories per year (or 2,046 calories per day) due to the computed costs of broad-spectrum plant gathering, which included walking roughly four hundred to five hundred hours a year and stooped/bent grass and plant foraging. In summary, hunters expended fewer work hours per year and benefited from higher protein; but broad-spectrum plant foragers and storers trended toward smaller body mass and work more hours annually, winning the human numbers game over time.

The Shift to Basketmaker Horticulture

With the onset of small-scale horticulture between 1000–500 BC, annual work hours and consumed calories rose among the gardener-foragers. Such families had a longer work year but could feed more children and store more food than strict families. Over time, the hard work of seed-dependent family groups created a modest power phase that added up to eight hundred years of increasingly sophisticated foraging strategies, thousands of planting experiments, cultivating, watering, and storage techniques. Thus, the transition to the vibrant, complex, and long-lived Basketmaker era took about six centuries. Without the women's early gardening successes, the Archaic way of life might have continued to dominate the San Juan Basin unabated for another millennium.

Gardening Progress in Basketmaker Society, 500 BC–500 AD

New strains of corn races had arrived from the Tucson Basin district and northwestern Chihuahua. From these strains, Basketmaker-era gardeners experimented and created regional races of corn that could produce a decent harvest in the lean soils and iffy precipitation of the San Juan Basin. These new local strains were easier to grind, an important efficiency, as corn grinding on a stone *mano* and *metate* was very intensive work. Some were superior in drought years; others produced ripe cobs in a short growing season. Several small varieties weathered cool nighttime temperatures in upland areas; still other varieties eventually developed deep tap roots (racemes), and could be planted in the sandy soils that overlay many clay deposits.[3] Corn gardeners carefully stored the most salutary varieties of seed corn apart from other varieties, with the hope that the carefully curated kernels would breed true.

   Best of all, corncob size and sweetness significantly increased thanks to crossbreeding efforts. Cobs were now doubled in length, measuring about four to five inches long, and each ear sported between eight rows—a variety called *maiz de ocho*—and sixteen rows of softer kernels. These larger cob varieties averaged about seventy-five calories, as compared to the thirty-five calories provided by early chapalote corn.

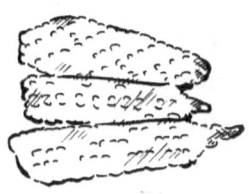

Teosinte, chapalote, and *maiz de ocho* sketch by Esther Burton.

Maiz de ocho corn. Drawing by Baker Morrow.

In due course, corn, beans, and squash or gourds were planted together, heralding a major step forward in garden health and productivity (more on the important relationship of corns, beans, and squash later). The larger-cobbed corn, squashes, beans, and gourds were harvested, processed, and dried for storage. Macerated squash meat could be dried into flat lumps or thin squash leather. This could later be rehydrated and cooked on a stone griddle as a squash and corn gruel.

There is no doubt that Archaic women's pocket gardens contributed to this major cultural shift to a modernizing Basketmaker identity in the northern San Juan Basin as well as parts of Colorado, Utah, and Arizona. Archaeologists and paleo-botanists have been able to identify many varieties of corn from their ancient garden plots. Many gardens contained darker soil, a "tell" of repeated bean plantings. These valuable varieties of corn and beans contributed to the best gardens, along with "cushaws," sunflowers, and amaranth.[4] In short, early women's pocket gardens provided enough diverse, high-quality food to elevate gardening to a relatively stable core lifeway.

Foreshadowing the Future

By roughly 300 BC the Chacoan world was still eight hundred to nine hundred years away, yet the shift to horticulture in the San Juan Basin had already introduced new organizing forces based on human ecological interventions. The consequences of these successes seem to have morphed into an illusion of control over nature among the most successful planters. Those planters were sited on good soils and reliable ground water. But those with no ground water could not match their more successful neighbors. This fact alone created a de facto class system.

Most hand-planted gardens relied on annual precipitation. In the somewhat wetter climate of about 700 BC to 200 AD, those labor investments paid off often enough to encourage a family's overplanting which, in good years, diminished the fear of famine.

If a crop failed, gardeners would plant extra corn rows the next year, store more than enough food for a year, and would *not* eat the seed corn. The vagaries of summer rains encouraged the addition of drought resistant crops like sunflowers, which also drew natural pollinators; thus, some districts' trends consisted of planting mixed gardens and *not* weeding out useful wild plants. The final cultural rule was to share with friends and families, whose gardens were in other favored locales, as part of their shared food networks. This was an ecologically sensible practice.

Ethnographic Trends

We know from worldwide ethnographic studies that many surviving forager societies shifted to gardening under pressures of modern population increases and commercial forest logging. In almost every case, the preponderance of gardening labor was provided by women and children. Female horticultural labor allowed for the care of children at a garden's worksite. Young moms could pause their weeding or watering to nurse an infant while on the job. This was likely the case during the Late Archaic and dominant early Basketmaker society.

Versions of this shift to gardening still exist in many less modernized parts of our current world. I vividly recall my hillside views of one southern Ecuadorian plantation, as hundreds of Quechua/Quichua-speaking women moved rhythmically through the long rows of daisy-like flowers harvested to make insecticides. About a third of one large group of several hundred women had either nursing babies wrapped onto their backs, or infants strapped into large baskets. The hacienda owner's overseers called *mayorales* followed on horseback, their ominous whips cracking whenever the harvesters slowed down.

Those mayorales artificially sped up the harvesting, which created a transitory power phase, rather like the enforced speed-ups of modern Midwestern meat-packing plants during the United States' 2020 COVID-19 outbreak. From the hacienda owner's perspective, the pace set by his whip-snapping horsemen generated efficiency, which enhanced his profit. In contrast, those same Indigenous women also tended their own small family food gardens near their mud houses, in hillside villages scattered along dusty dirt roads that had once been part of the Inca road system.[5]

But without the whip-bearing mayorales pressing them on, their own diverse gardens produced more food and crop per quarter to half-acre home plot than did the elite's vast plantations. The small home plots were tended more carefully, weeded more frequently, received more urine and feces (from humans, children, guinea pigs, wild cats, dogs, chickens, etc.) in a smaller area—which meant a touch more nitrogen in the soil—better weeding, and hand-watering as needed. This conforms to the same pattern of land use at the seventeenth-century pueblos of Abo and Quarai in central New Mexico. Those pueblo farmers' smaller garden plots out-produced the Catholic priests' large, Native-tended gardens.

Steps in the Long Path to Sustainable Gardens

During the Late Basketmaker Period of about 300 AD, the cultural underpinnings of what would become the Chacoan World in just another five hundred years had begun to emerge. The fundamental dynamics of this period include the following:

1. The regional cultural attraction of a now-stable Basketmaker society.
2. Continual experimentation with strategies to grow corn stalks that matured quickly, were drought resistant, and produced more cobs of larger corn. By the Common era (AD 1), new and hybrid

corn varieties produced up to *triple* the calories per cob compared to the early chapalote corn. This near tripling was a huge efficiency gain that appears to have supported both larger families and larger storage pits.

3. More labor investments in garden plots added useful wild plants, yuccas, piñón nuts, prickly pear, currants, wolfberry, sunflowers, and semidomesticated plants like amaranth, and evolving varieties of corn, common beans, squashes, and a variety of gourds.

4. Experiments with landscape infrastructure such as hand-placed stone alignments to slow and direct hillside runoff from the slopes above the gardens, larger bell-shaped seed storage pits less likely to cave in, and experiments in small-scale landscape modifications to direct the flow of creek water to nearby gardens. By 500 BC, low brush and cobble borders popped up everywhere that gardening thrived. Tweaking the landscape enhanced crop yields.

5. Evolution of lineage-based social organizations led to claims of "absolute ownership" of carefully sited garden plots. Since garden ownership was female based, women's status rose while opportunities for enhancing male status lagged.

6. An increase in innovations: use of dried gourds as containers and larger stone grinders. Among the gardening tools were hoes, tri-faced stone soil breakers, fire-hardened digging sticks, rabbit sticks, pitch-tarred baskets to carry water, well-made yucca carrying baskets, sandals, and refinements to woven nets (of yucca fiber or human hair) used to catch rabbits.

The Power Phase of 500 BC–200 AD

The period from 500 BC to 200 AD was the fastest paced, most labor intense, experimental, and socially pioneering era in the San Juan Basin's seven thousand preceding years. It was revolutionary, a genuine power phase—labor inputs rose rapidly, and society complexified by leaps and bounds.[6]

Occasional surpluses of corn and other dried foods provided both new opportunities for trade and increased risks of raid. The side effects included increased security costs to protect gardeners and their stored crops, more complex trade networks, and the rising eminence of farmers sited on permanent water.[7]

To expend the labor calories required to create gardens during this experimental phase, human body size and life span likely declined a bit. Borrowing against height, robustness, and longevity is a biological solution to relentless manual work pressures. Those pressures imposed heavy workloads among women and men of that era. The daily food calories they needed to fuel their hand gardening work rose dramatically. Thus, the larger cobbed corn being grown provided a huge and immediate efficiency gain to many gardeners at this time.

As new and younger families were expanding into the stingier wild food districts of the unsettled San Juan Basin, the ratio of gardeners to foragers rose rapidly. The best gardens located on ground water with good soils had long since become prime properties. New reliable gardens were sited in mid-elevations where the key to survival demanded an increase in stored food. Carefully guarded seed corn was planted every year in spite of frequent droughts; therefore, the wetland farms were held tenaciously and protected by larger family groups. These produced a class of regional "market makers."

Annual labor costs among gardeners skyrocketed. High-yield gardens required careful weeding, thinning out weak stalks, and carrying large gourds of water to get cornstalks through a dry summer. These labors elevated daily workloads and engaged in the long scramble to grow higher-yielding crops. Each well-watered diverse garden plot became its own human created ecotone. The suite of crops grown in that plot and hundreds of others reduced native niacin-rich plants to produce even more corn calories.

In the long term, the rise of corn and decline of niacin-producing plants created a hidden and misunderstood barrier to regional health, even though it offered an immediate boon to regional markets. Corn offered low protein content, even though its carbohydrates provided a temporary sense of fullness, dulling frequent pangs of hunger. Its best dietary contribution, however, was the discovery that its kernels popped and fluffed up when heated over a fire. The carbs and satisfying popcorn fluff were all enough to merit it space in ancient gardens and beyond.

Overall, the dietary benefits of early corn varieties was middling. It most likely became *the* staple food because of the sheer food volume it supplied as the San Juan Basin's population rapidly increased between 500 BC and 400 AD. If just one additional healthy baby was born each year to a population of only one hundred farm families in a lower elevation corner of the San Juan Basin, their local population could have risen to almost nineteen hundred people in just three hundred years! That probably did not happen, given infant mortality, niacin deficits, and over-foraged landscapes.

By the dawn of the Christian calendar, there was even more pressure to increase crop yields and secure land holdings. A tenfold increase in population—or even half that growth—would have dictated larger gardens, higher labor inputs, a continual frenzy of corncob selection to achieve larger cobs,

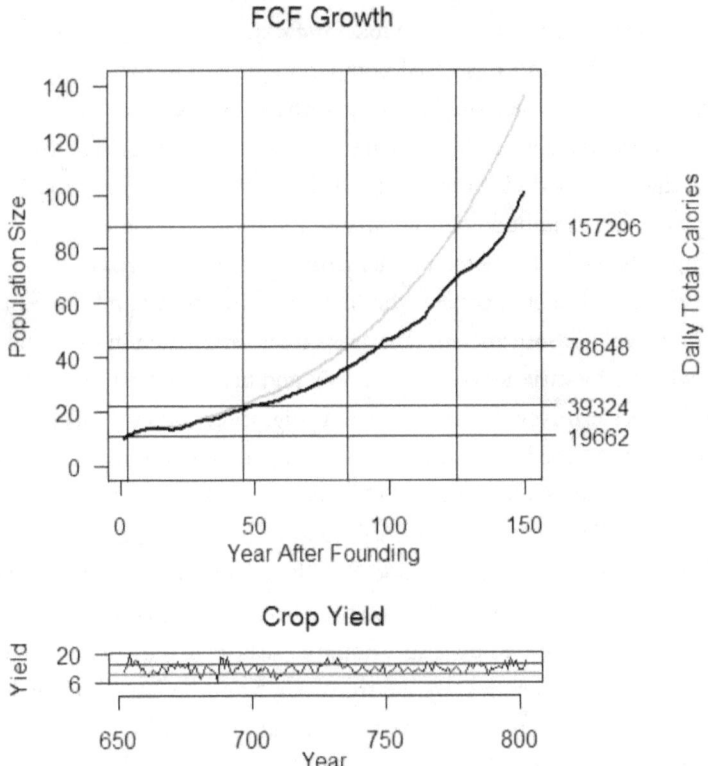

Figure 3. Four Corners Family (FCF) Growth.

softer corn that was more easily ground, and cornstalks that produced more cobs than the two-cob-per-plant versions of the earlier corn varieties.

Over time, the annual efforts to expand gardens began to pay off. That encouraged even larger gardens and a slowly diminishing reliance on foraged wild plants in some of the drier locales. The long upward trend in population resulted in more intense labor, more domesticated crops, and expanding garden size to feed this era's multigenerational power phase.

Clear evidence of that energetic trend can be found in the simple but important charts of archaeological site numbers in New Mexico's San Juan Basin. In Timothy Kearns's chapter of editor Bradley J. Vierra's *The Archaic Southwest*, he lists the number of archaeological sites by time period from the Middle Archaic to the Late Archaic, a period by which the number of identifiable archaeological sites had increased *eightfold* in a thousand years (Kearns 2018). This was *the* growth rate that triggered the Chacoan world.

In the second edition of my book, *Anasazi America*, my former student Christine DuBois and I assessed population growth in the San Juan Basin between the end of the Late Archaic and the peak of Chaco Canyon society at about 1100 AD. DuBois's summary calculates that population grew *ninefold* in that time period (Stuart 2014). It is satisfying that Timothy Kearns, using a different methodology, computed a similar increase.

Linguists have also detected elements of ancient language patterns in historic period Pueblo peoples who later lived in and around the San Juan Basin. Several of those ancient languages likely originated in the same areas from which new varieties of corn, squash, and beans arrived in eastern Arizona and western New Mexico. Both in-migration and long-distance trading patterns enriched the Basin's mix of languages and human DNA. New Mexico provided much of the early "Aztec" turquoise.

While the San Juan Basin was still food-rich in the high country during the Middle Archaic Period (ca 3600 BC), its peoples were not isolated from a larger world. Rather, they enhanced their own world by absorbing new crop varieties, horticultural knowledge, Sonoran seashells, human genetic material, and linguistic changes. By the transition from the Late Archaic Period to Basketmaker culture, newcomers of the San Juan Basin likely spoke several different languages.[8]

All of these factors—innovation, imports, planting techniques, small-scale irrigation, and the arrival of newcomers—contributed to the pace and directionality of change during the power phase of 500 BC to 200 AD. Technological trends combined power enhancing factors (population growth and more work hour inputs) and new efficiencies (fewer food calories thanks to better storage techniques). Corncobs increased in size, in number per stalk, and tripled in calories. The region had come a long way from earlier foragers' modest, abandoned dugout shelters of 500 BC, with disturbed soils filled with amaranth plants.

FARMING AND POPULATION GROWTH

Having babies has always been "expensive." In the prehistoric Southwest between 500 and 500 CE, the total dietary cost of a pregnancy and nursing for two years was roughly 250,000 to 260,000 calories beyond normal female dietary needs. That computes to roughly four thousand ears of moderate-sized prehistoric corn (at sixty-five calories per cob) or about one-eighth acre of hand-cultivated corn annually. Doubling the number of our Four Corners family in forty-eight years, given that only an estimated 65 percent of pregnancies produced a live child at age one, required fifteen live births, or four million additional dietary calories. This required a notable increase in the acreage that each family had to farm over time.

Source: (Stuart 2014)

CHAPTER 5

FARMING LABOR PATTERNS

BY THE START OF the Christian calendar in AD 1, in the era archaeologists refer to as Basketmaker, many of the San Juan Basin's women and girls worked harder than their great-great-great foraging grandmothers (Crown and Fish 1996). Farm labor was more intense than foraging. Such small-farm families continued to rely on seasonal foraging, resulting in a de facto two-pronged food economy.[1] In a late summer season when abundant monsoon rains materialized, most gardeners could produce ample crops. With a substantial crop of corn, food storage made those farms workable from year to year *if* the highest yielding crop varieties could be harvested and successfully stored. This involved thoroughly dried corncobs and clean grassy pit linings.

During the summer monsoons between 600 BC and AD 1, the elements

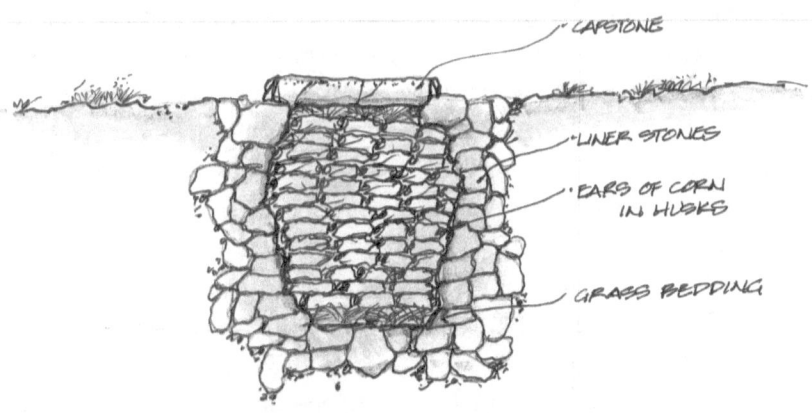

Corn storage bunker section. Drawing by Baker Morrow.

of a completely balanced meal of corn, beans, and squash could reliably be grown in a creek-side garden. If that creek or wash flooded its banks, those gardens could grow in size over time. According to noted scholars R. Gwinn Vivian and Adam S. Watson, these locales near flowing water often produced surprisingly large corn crops (Vivian and Watson 2015). Supporting evidence of this is based on analysis of late nineteenth- and early twentieth-century Diné (Navajo) corn planting successes along the canyon washes near Chaco Canyon's Pueblo Bonito (Heitman and Geib 2015). Excess corn could be traded for meat, other foraged wild plants, grinding stones, obsidian (dart heads and knives), etc. Most families sited on these favored farms could support larger than average families *if* the late-summer rains came. Women could retain their claim to both foraging rights and to owning and controlling their gardens. The basis for this assumption lies in the huge body of ethnographic literature on hunter-gatherers. As will be mentioned later, forager groups tended to transform their food culture when facing high population density on once thinly populated land. Many transformed into *gardener-hunter* bands in which women worked and controlled the actual gardens. In those societies, men continued to hunt or forage in nearby forests, but the women typically produced, processed, and controlled the garden-based food supply (Lee and DeVore 1968).

Similar trends appear to have cemented farming women's rights to pass on their gardens, creating something akin to a "real estate" asset based on rights of first use that passed gardens to daughters and granddaughters.[2] These practices tied some group members to the best farm sites but also freed at least some young men to trade, hunt, or to work on infrastructure—such as building larger and more comfortable family shelters, digging storage pits, cutting, and transporting heavy shelter beams, firewood, or fashioning stone garden windbreaks and edging. Girls began to help with cooking by about age eight, and by about age twelve became the primary corn grinders (Crown 2000a and b).

A rough estimate of a young adult gardening woman's annual labor in the San Juan Basin between 600 BC and AD 1 suggests about 720 annual hours of moderately intensive garden work averaged roughly 300 calories per hour (food-producing work calories), plus about 1,000–1,100 annual hours of combined childcare, body care, basketry, sandal making, clothing, weaving, foraging for yucca and grass fibers, and food preparation at 150–170 calories per hour.[3] In contrast, cornmeal grinding with *mano* and *metate* is estimated to cost about 400 calories per hour. Water carrying was primarily a young woman's task; water needs were often delegated to girls in their early teens. In short, a young proto-gardener woman's work year between 500 BC and AD 1 likely consisted of at least 1,400 work hours. Adding in shelter wall maintenance, fire tending, and harvest season food processing, a year's workload for a female climbed to about 1,800 hours. There would have been time for leisure after fall harvest and less work in midwinter.

In contrast, the standard work year among traditional hunter-gatherers was defined by Richard Lee and Irven DeVore's classic ethnographic studies in *Man the Hunter* to be about five hundred hours of active work annually

CARRYING WATER—YOUNG WOMEN'S WORK

Unmarried teen women without children were more metabolically efficient water carriers than were young mothers who also had to carry their babies. This implies that most water carriers were in the 11–12 to 16–17 year age group. Assume a robust, 110 pound female in her late teens. She carries 2.5 gallons of water at a pace of 3 mph. She burns 97 calories an hour and needs 2 or 3 full corncobs to refuel. At a walking speed of 3.5 mph, she burns 112 calories, or 3 full corncobs of fuel. Total weight of 2.5 gallons of water is 21 pounds. Average distance to water, derived from archaeological field notes, is half a mile or less. Derived assumption: Since a young woman with a nursing infant strapped into a wood-framed back cradle would have added 12 pounds of baggage, that would have amplified water costs by about 65 calories per hour. Note that given a high carbohydrate but low protein and vitamin diet, most young women of that era weighed less than 110 pounds, and would not menstruate before age 16.

(Lee and DeVore 1968). Among women in the San Juan Basin, the work year was at least double that of their ancient grandmothers of the Middle Archaic (about 1200 BC).

As an ethnographic generality, men's work in classic small gardening society included hunting as well as the creation of infrastructure, including digging the shallow, dug-out pithouse floors, creating garden borders, small-scale creek-side irrigation channels, wood cutting for the frames of campsite shelters, storage pit digging, stone tool making, and atl-atl production. They likely also served as short-term help in the gardens, breaking soil with their fire-hardened digging sticks to plant seeds, then, in fall, with hands-on help at harvest time. Gardening emergencies, like an early fall snow, likely pulled in every able-bodied person to save a crop.

Estimates of the male work hours rest on much more of a guess: let us make it a cautious fifteen hundred hours annually. If we add in the night hours of "security watchmen" during the harvest and storage season, we can round up to about sixteen hundred hours a year. Yet in just a few more centuries, these would likely be remembered as "light" work years.

A New Millennium Arrives and a Regional Society Advances

By AD 1, the better-watered districts of the San Juan Basin had become economic and demographic magnets, drawing gardeners to higher food-producing and increasingly crowded territories. Food calories and protein produced in smaller spaces was efficient. As farmsteads became larger and shelters more ample and complex, it made no sense for the whole family to pack up and forage in the uplands every fall, particularly if the garden was located on or near reliable water sources. Year-round residency in order to protect the rights to valuable wetland gardens likely became common as the central San Juan Basin became more crowded.

Large forage parties still deployed each fall in search of deer, acorns, and piñon nuts. Indian ricegrass in particular was still an essential early spring food source. Some family members likely traveled to their ancestors' upland camps to casually tend them, as at Jemez Cave. This may have served to maintain a family lineage's upland foraging rights to fall camps created by earlier generations of their own family. Though few scholars have addressed

TABLE 4. Caloric Costs of One Work Hour—Gardening

Activity	Metabolic intensity scale	Calories/hour	No. of cobs
Clearing light brush	3.5	459	6
Weeding and cultivating	3.5	459	6
Planting and stooping	4.3	564	7.4
Aggressively picking fruits and vegetables	4.5	591	7.7
Weeding and cultivating with hoe or digging stick	5	656	8.6
Laying cobbled rock	6.3	827	10.8

Assumptions: Men assisted with digging sticks and planting. They also likely added cobbled "walls" to enhance water flow and remove large stones that impeded planting. Source: captaincalculator.com

this phenomenon, it likely played a role in the cultural ethos and enhanced the social complexity of the future Chacoan world—still eight centuries away.

The Difference of a Thousand Years

Between about 500 BC and 400 AD, many of the basic cultural, socioeconomic, and technological elements that would later characterize the Chacoan world were in play: successful gardening, effective food storage, seasonal landscape usage, more complex infrastructure, increased work hours, larger gardens, better shelters, water control practices, and expanded trade with districts in Arizona, California, and northwest Mexico.

Work hours rose for nearly everyone in the San Juan Basin. In the words of Timothy Kearns (2018), "The increased number of sites generated an apparent increase in occupational intensity around 2,400 years ago (about 400 BC), and coincides with the common occurrence of 'farming' sites in the San Juan Basin." Initially, female lives changed the most. The female leisure time once experienced by broad-spectrum foragers diminished rapidly as work in the gardens became both harder and longer. Breaking soil and hoeing with digging sticks, antler mattocks, and bone rakes consumed an enormous amount of work calories, as did stooping to hand plant dried corn kernels in each hole. For gardeners who did not have garden plots on naturally watered locations, there was the calorically costly process of fetching water during summer's dry periods. The region's earliest clay pots came about 350–400 AD, a boon and efficiency for the water girls. Among the gardeners, men's workload initially rose more slowly than the women's. That, too, would change as San Juan Basin populations grew in number and began to stratify into haves, have-nots, and a small but steadily expanding magico-religious class. By AD 1, rapid change during the prior five centuries had reshaped a simpler world into a rapidly complexifying one.

Visions of a New Millennium

By AD 1, thousands of garden settlements dotted the San Juan Basin. These sites produced corn, gourds or squash, common beans, as well as crops of friendly wild invasives like amaranth, leafy chenopods, and sunflowers. By square footage, those gardens produced more calories and protein per labor calorie invested than any of the natural plant collecting niches, except piñon nuts.[4] Upland ecotonal zones (sixty-two hundred to seventy-five hundred feet in elevation) offered pockets of rich dropseed grasslands that abutted piñon-juniper stands. Such econiches provided protein-rich nuts and acidic juniper berries. The acidic juniper increased protein uptake from a meal of corn during human digestion. These grassy uplands were also species rich in rabbits, wild turkey, deer, or an occasional elk grazing in late season grasses. The downside of those higher elevations was the short growing season and small-cobbed corn.

Gardens sited on loamy hillside soils near water sources out-produced most other farming niches. The best field archaeologists of the 1970s to 1990s always noted "distance to nearest water" on their officially filed site forms in Santa Fe's ARMS file. Carefully placed cobble rows could divert a hillside's runoff to a specific corn plot, dramatically raising its yield.

In dry years, fall and spring plant foraging was frantic. Belonging to an ancient lineage "sharing network" dramatically reduced a family's starvation risks. Broad-spectrum forager groups likely cooperated with nearby corn growing families to efficiently harvest large patches of spring's Indian ricegrasses in return for a share of the seeds.

Relentless increases in regional population had diminished the hunters' returns on big game. As a consequence, meat protein increasingly came from smaller, more numerous animals like rabbits, prairie dogs, box turtles, rats, collared lizards, and occasional wild turkeys. Though some partially excavated sites of the late BC and early AD era have yielded bones of bison, elk, deer, bighorn sheep, desert sheep, and bear, these foods had become quite scarce by 500 BC. Many archaeologically recovered large-game animals appear to have been "parted out" in food exchanges. At the same time, unpredictable precipitation and wild temperature swings frequently shaped availability of both wild food and farm produce. Over the next five centuries, erratic rainfall, diminishing plant foraging zones, and increasing population would shape a *less* diverse ecological world in the San Juan Basin.

CHAPTER 6

THE EVOLUTIONARY BALANCE SCALES SHIFT, 1-600 AD

THE PERIOD BETWEEN AD 1 and 600 AD produced more cultural changes than had the previous twelve hundred years. Thermodynamically, the balance scales moved to a more frenetic rhythm. The prior twelve hundred years had been largely devoted to experimenting and obtaining more plant-based calories, as well as increasing diet diversity produced in women's pocket gardens. Going forward, the San Juan Basin would focus on producing ever more food, as well as making technological innovations in water distribution to support gardening for an increasing population.

The basics of food security increasingly depended on planting drought-resistant corn and recruiting several distant family trading partners to one's sharing networks. The sharing networks might have worked indefinitely if population growth had been curbed in this era. But it was not! Successful gardening and corn farming communities produced more babies, which shaped a long era of food struggles among small farm populations. The solution was a world in which humans took control of the land and reshaped it to meet their needs. Yet, this new world did not yet include any effective regional strategy to overcome the mysteries of unpredictable precipitation. Thus, expanded storage space within early shallow, lobed pithouses and larger, more numerous hillside storage pits became important visible signals of increasingly intense horticulture.[1]

Overview of Life Changes in the San Juan Basin

There do not exist enough data points to reconstruct a refined portrait of the many changes during this era. Yet between AD 1 and 600 AD, site records help us deduce the overall cultural trends:

- More goods were traded like early brownware pottery, turquoise, tool materials, shell from the Pacific coast, bone adornments, turquoise, and seed corn.
- Declining protein intake among corn growers and their narrow-spectrum diet diminished regional health.
- Population growth was robust, and required more food, larger gardens, increased food storage and security, and new ways to bring water to crops.
- Sharing networks became more widely used as a response to continued spotty and unpredictable rainfall, gaining economic power for a time.
- The hidden genetics of seed corn's distinctive varieties—drought resistant, large cobbed, etc.—were slowly uncovered.
- Dried corn husks and cobs became sources of fuel, as wood became scarce in lowlands.
- The unstable climatological factors of garden and corn crop success led many gardeners to produce larger crops than needed in any one season. Storage and water technology advanced, and regional corn markets emerged.

Successful Corn Experimentation

The first six hundred years of the Christian calendar era saw a much increased population in the San Juan Basin primarily because gardeners' experimentation with larger cobbed and easier-to-grind corn had paid off. Work calories invested in a 500 AD garden one eighth of an acre in size produced at least twice the edible corncobs compared to harvested stalks of 500 BC.

If left standing during the winter, the dried cornstalks trapped drifting snow and drew rabbits, wild turkeys, and migratory birds, all of which

STORAGE PRACTICES

The emerging thermodynamic efficiencies of more intensive gardening and storage practices that emerged by roughly 500 BC, a slightly wetter period, had created the caloric platform for a larger, more innovative population. The reward was both more living human bodies to feed and many more human work calories demanded per day than were required of their ancestors. The benefits of early food and food processing efficiencies morphed quickly into a new realm characterized by much harder work and the increasing risks of regional starvation when the late-summer monsoon rains did not materialize.

were competing for missed corn kernels. Foraging coyotes, weasels, rabbits, and deer were also attracted to fields with standing cornstalks. Their collective nibbling, excrement, and rooting in the dry cornstalk patches fertilized future plantings. Surrounding juniper trees dropped prolific acidic needles and berries in the garden, which reduced soil pH levels and improved tilth.

Basketmaker II Culture: A Look Through the Rear Mirror

Archaeologists typify this period as part of the early Basketmaker era, which flourished by about 500 BC and was modified enough by 500–600 AD to be labeled as Basketmaker II. The classic archaeological descriptions of this cultural phase focus on the following:

- The presence of corn in most sites
- Use of large, carefully woven utility baskets, ranging from plain to elaborate
- Efficient brush-roofed or cave-sheltered family camps
- Cliffside storage of dried corn, both secure and efficient
- The replacement of atl-atls with bow and arrow

- The appearance of simple, coiled grayware pottery that was efficient, utilitarian, and had economic trade value

Tools of all kinds were being refined, corn had become a basic food, as had squash and, to a lesser extent, beans. Extended-stay family camps were large and multi-seasonal—far more durable than the short-season camps of earlier eras.

Analysis of camp floor soils by the flotation process indicate that both domesticated and foraged foods were still consumed. The small creatures of yore that had provided both protein and fat to the ancient "critter fritters" were still being eaten, but analysis of toasting stones suggests that corn, squash seeds, amaranth and, in places, piñón nuts, had become primary sources of protein. This is confirmed because toasting stones sometimes trapped microscopic food traces in the pores of their cooking surface. Most of these stones were squarish to oblong, made of grainy basalt, and about as thick as a modern brick. They would be preheated in a fire to red hot, then ground corn meal, amaranth, ricegrass, sunflower seeds, etc. would be added and mixed with mashed, protein-rich small creatures. The stones' mass absorbed enough heat to griddle cook a meal. That practice left an inventory of tiny, partially cooked seeds and grasses for archaeologists to analyze.

Opportunistic food intake was more or less a continual daily process, as people chewed and sucked on sugar and carbohydrate-rich grass stalks

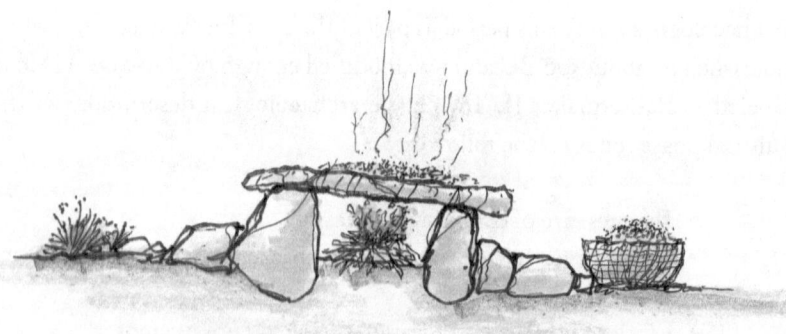

Comal, toasting stone, with grains and a basket. Drawing by Baker Morrow.

throughout the day. Those small carb and sugar hits were used just like a "Coca Cola breakfast" in low-income American families in Appalachia and the deep South right up into the 1960s–1970s.

The Basketmaker era's cliffside storage niches in the Mesa Verde district were easy to protect, but the smaller single-family camps were harder to secure. Thus, the trend in some districts shifted to larger, more permanent camps on mesa tops or hillsides. An isolated family's security was a serious issue. In many districts, solitary homesteads were replaced by clusters of families. Some of these clusters gradually expanded into villages.

Gardening Efficiencies

Gardening efficiencies continued to be refined during the Basketmaker era, yet dryland farmers remained vulnerable to unpredictable rains and droughts. Even if their gardens and corn plots were sited on nutrient rich soils, their crop's fate rested on the volume of summer monsoons or hand-carried water. Water jars made it much easier to carry water and eventually led to new gardens sited at further distances from ground water. Dryland farmers learned to store large quantities of food, and corn was the easiest to store.

Meanwhile, gardening communities in the high country's forest regions who also relied on seasonal rains and winter snows to produce small-cobbed corn had alternative food options to fall back on—large game, piñón nuts, juniper berries, and acorns. This rendered the most productive upland camps nearly as stable as the wet farm communities, *if* the winter storms yielded deep snowpacks.

Anatomy of Two Innovations, 500–700s AD

In order to understand the dynamics of the 500s–700s AD, one must focus on nutrition, corn, and crop efficiencies as the basis of cultural driving forces in the San Juan Basin. Though corn was the commonest human dietary staple in regional diets, it is impossible for the human gut to extract corn's stingy protein, called lysine, unless the corn is treated with lime rock or other sources of acidity, like juniper ash.[2] Corn consumed during the Basketmaker II era was likely cooked with a touch of lime-rich rock or other acidic media like juniper branches, ash, and berries. The acidity of those sources transformed more of the corn's lysine into a chemical state that a human gut could metabolize.

Happily, one other act of nature played a significant role in corn nutrition: corn fungus. When those purple-gray bulgy growths known widely as *huitlacoche* in the Aztec language (*nahuatl*) invaded corn plots from spores in the stalks' growing soil, corn crop yields were somewhat diminished, as many infected cobs would not fully mature. Fortunately, those lumpy, purple-gray, smut-infected cobs actually produced much higher levels of metabolizable lysine than uninfected corn. The typical protein content of corn varies, by both growing conditions and subspecies, from 3–6 percent protein, but the protein content of *huitlacoche* cobs ranges from about 9–18 percent protein (Battillo 2018).

If the various peoples who inhabited the San Juan Basin in the 500s AD could have foreseen events of the late 700s AD, they would likely have held a tighter rein on traditional, culture-based population controls but, like all humans, the future was a mystery to them, and the dynamics of misjudgments were obscure when viewed through the complex lens of human imagination and hubris. Those who shout gleefully, "We got this!" often fail to realize that what may be under control one day could well blow up the next. All human behavioral trends are both malleable and short-term when viewed on the vast time scale of evolution.

Bow and Arrow as Power and Efficiency

Some human innovations do have long-term impacts. Take, for example, the bow and arrow, which became a hunting tool and main source of defense by men and women while slowly replacing the atl-atl. The bow and arrow came to the San Juan Basin about 300–500 AD.

The bow and arrow's technology offered several advantages over the atl-atl. An archer could nock and launch arrows at more than twice the rate of a hand-thrown dart/spear. The bow also favored many men and women in a way the atl-atl did not—long arms and short arms became more equal in actual hunting, just as Samuel Colt's model 1873 six-shooter rendered a small man able to hold his own if threatened by stronger, bigger ones. The nineteenth-century American frontier West universally referred to Colt and Remington's six-shooters as "equalizers."

So were the bow and arrow. The bow's arrows averaged roughly two to three feet in length—much shorter than the five-to-six feet of an atl-atl dart. Petite females might have struggled with the atl-atl, but they had no trouble with bow and arrow—their shorter arms were an advantage. The arrow's stone tips also weighed less. Aided by a twenty-five pound bow, the arrows flew at a greater speed than an atl-atl at about three hundred feet per second—thus, they penetrated quite deeply. Precious obsidian or other flaked rock traded into the San Juan Basin from the Jemez Caldera to the east could produce two to three arrowheads from the same sized obsidian blank that had been necessary to produce one heavy atl-atl dart. A three-to-one efficiency of material consumption!

Finally, the bow and arrow required less hunting skill. The arrows flew well over a hundred yards, reducing the agonizingly slow and quiet sneak-up phase of an atl-atl hunt. The bow was faster to reload, more efficient, had somewhat greater range, and reduced the chance of a wounded animal's horns winding up embedded in a careless hunter's midsection. It took only two weeks to a month of regular practice to become reasonably accurate with a bow.

In contrast, it took two to three *years* of practice to become proficient and accurate with an atl-atl. In addition, Paul Reed and Phil Geib point out (2013) that a bow could be fired from a confined or hidden position. In contrast, an atl-atl required a fully exposed stand-up position. We are talking hunting efficiencies here, where velocity doubled while the mass of the projectile declined. Interestingly, in the early days of the bow and arrow, about 350–500 AD, the atl-atl-using Basketmaker regions north and west of the San Juan Basin had clashed more often and more lethally than bow users in the basin's late Basketmaker period. Had the bow's equalizer effect also enhanced behavioral caution in the San Juan Basin?

Gardening Expansion

The arrival of the bow and arrow may have further reduced the availability of large game in the San Juan Basin during the late Basketmaker era. That might have played a role in gardening expansion. The nature of horticultural enterprises likely took several forms. The first of these was an increase in both the size and number of established gardens. Garden expansions were most easily accomplished on prime ground-watered plots. They would also come to play a singular role in the continuing transformation of San Juan Basin society. But planting expansions often required the creation of small, rock-lined, creek-side irrigation ditches. This infrastructure was calorically costly. Moving rock and reshaping the banks of creeks or intermittent washes burned about four hundred food calories per hour. Among late Basketmaker and early Unit Pueblo homesteaders, the work efforts of landscape shaping and water control often exceeded a small community's capacity to both garden and engineer landscape improvements. Ancient sharing networks may have assisted in the transformation to larger gardens that supported larger families.

Thus, during the 300s–600s AD, expensive infrastructure needed to move water combined with the costs of building more living facilities. Both gardens and dwellings became more expensive and more complex, stressing available labor, food costs, and increasingly scarce timber for building.

In dryland garden districts, agricultural expansion was slow. As sunny hillside hamlets consisting of three to six families proliferated in the southern and central San Juan Basin, pocket gardens continued to be planted on isolated hillocks and carefully tended by women, providing crucial food diversity: specialty corn, squashes, beans, and semidomesticates like amaranth, sunflower, chokecherries, or possibly a wolfberry bush.[3] A bit farther downhill, however, new and larger gardens planted corn, squash, and beans.

By this time, it had become practice to shift corn rows every year. This prevented entire cornfields from having to lie fallow—instead, new rows of corn were shifted three to five feet from the prior year's planting rows. Beans and squash were then planted in the fallow rows. This practice has often been overlooked by scholars, who compute the corn harvested based on an estimate of the total size of a field. Those big fields instead need to be computed at about 50 percent of gross square footage in production in a given year.

A closer look shows that this cultivar mix of corn, beans, and squash represented the apogee of a nutritionally effective small-garden crop technology. It is another example of the supportive "reciprocity" between plants that

SHARING NETWORKS AS LIVING CELLS, PER THE SANTA FE INSTITUTE'S BRAIN TRUST

According to Stuart Kauffman, living cells send out chemical "messengers" to trigger the development of other cells in an embryo in a self-consistent network, instead of just a lump of protoplasm. This concept resonates with William B. Arthur's ideas on the self-consistent, mutually supportive "webs" of interactions in human societies.

Kauffman's ideas heavily influenced Arthur at the Santa Fe Institute. In fact, Kauffman inadvertently describes the dynamics of the ancient family sharing networks, which I assume generated social and economic information mechanisms in foraging society. Those mechanisms established social order over wide geographic districts. In short, each new "cell" was a new family added. Each new family addition expanded both the network's labor pool and its territorial expanse. The developing network cells in turn acted as a human agent of negentropy: order, calm, rescue, and a font of regional information. As new members of the networks increased, the information of the earliest ancient members and their rules remained together.

Source: (Waldrop 1992)

Corn, Beans, and Squash

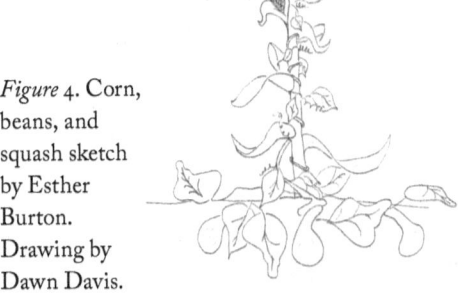
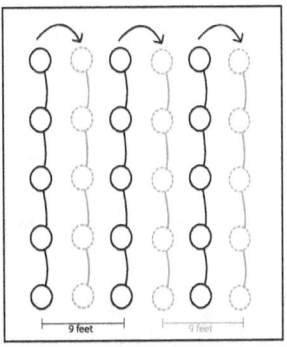

Figure 4. Corn, beans, and squash sketch by Esther Burton. Drawing by Dawn Davis.

Arrows show rows' alternate-year spacing.

is elegantly discussed by Robin Wall Kimmerer in her book *Braiding Sweetgrass: Indigenous Wisdom, Scientific Knowledge, and the Teachings of Plants* (2013). In spring, the corn pushed up small stalks that young bean tendrils climbed. Beans added crucial nitrogen back into the fallow soil. The maturing squash leaves provided dense shade for the growing corn and beans and helped the ground to retain moisture, reducing (though not eliminating) the need for tedious hand-watering.

Garden fertilizer was the result of another symbiotic exchange between plant and animal life. By midsummer, many gardens had received human, dog, and rabbit excretions. The early Basketmakers and Puebloans likely made "toilette" in the privacy of tall rows of tasseling corn.[4] Pollinators such as birds, bees, butterflies, and small animals were also essential to a thriving corn plot.

While elements of this regional dietary strategy of corns, bean, and squash were likely already more than a thousand years old, it reached a new peak during the San Juan Basin's late Basketmaker II period (100 BC–500 AD). By growing these three crops together and allowing wild plants like amaranth and sunflowers to encircle the pocket gardens, women gardeners had created a miniature protein-rich, plant-based ecosystem in each garden.

Despite ample archaeological notes on corn, beans, and squash, there remains some mystery about beans. Botanical traces of beans have been hard to find in the residue from campsite soil samples. The jury is out on why beans are hard to document. Landscape scholar Baker Morrow points out that beans have small root systems that decompose quickly. My own suspicion is that some bean skins excreted in human coprolites had been eaten by ants, beetles, rabbits, and other rodents. Beans also tend to explode and shed their outer skins when heated in acidic water—thus, bean skins may have been disposed of in the cooking process. However, we do know from botanical studies that a number of bean varieties were available. But how often they were planted and successfully harvested in the San Juan Basin during the first few centuries AD remains unknown.

Preserving the Pocket Garden

Tending multi-crop gardens required a lot of labor and skill. This may be why the women's smaller pocket gardens received the most careful attention.

An acre of male-tended corn rows produced less protein than a third of an acre of corn, beans, and squash tended in the women's pocket gardens.

Pocket gardens used little space (two thousand to twenty-five hundred square feet was common in this time period), required less hand-carried water, could be managed by one or two people, and were impressive protein, fiber, and vitamin producers. They were also the de facto laboratories of many crop experiments. Such experiments came at fairly low labor and infrastructure costs. Many lessons learned in tending these pocket gardens wound up centuries later in large hillside and creek-side gardens tended by larger family groups.

The San Juan Basin's pioneering female gardeners had taken centuries to pass on key elements of their experiments to their descendants. Those experiments teased more bulk food harvests from dry, nitrogen-challenged soils. *Ancient women's invented southwestern garden ecosystems were as significant in its own era as the discovery of the genetic double helix that has shaped many new realms of modern knowledge.* That gardening genius was fortified by increasing access to squash and beans from southern Arizona by about 600 BC (Vierra 2018).

If female gardeners learned to create new, miniature ecosystems at will, they could improve their own lives and pass on the dietary benefits of their knowledge and handiwork to their children and grandchildren. Their diets increased longevity, raised their children's heights, and reduced failed pregnancies. Gardens that included high-protein amaranth elevated immune systems. Women gardeners could support a regionally growing population so long as more small and diverse gardens were planted.

Archaeologists do not know just how many of these culturally created miniature garden ecosystems were established in the San Juan Basin from 500 BC to 600 AD, but it was likely in the thousands, if not the tens of thousands. In any case, they were ecologically genius, as regional overreliance on one mediocre food crop, corn, for the bulk of its dietary calories was both environmentally and nutritionally risky over the long term.

Pottery and Cooking Efficiencies

In general, pottery ushered in a significant change in cooking techniques. Plain, local pottery graywares dominated in the San Juan Basin during the 300s–700s AD. This included medium-sized, gray-brown clay jars and pots, many with a rough bell-shape. The average pot held a quart or so. Gray utility wares were not surface polished or decorated beyond a neck coil. Most were used as cooking pots or storage jars. The ceramics of this period also had good trade value.

Prior to the arrival of pottery-making technology in the San Juan Basin, cooking had been done on spits (meats), slab-like cooking stones (critter fritters and/or grains and seeds), or in skin-lined stone boiling pits (corn/veggies/seed mash). The caloric dynamics of the change from stone boiling to fired clay cooking pots were analyzed by a team of four of my most gung-ho undergraduate students at the University of New Mexico: Mike Smith, Stella Kemper, Roy Huddleston, and Bradley T. Varner, along with the guidance of a modest, semiretired nuclear engineer (Stuart 2014). The results were fascinating. The new pottery-cooked food consumed *more* firewood than the ancient and tedious stone boiling method. An open fire under a cooking pot lost more heat calories to air than in the stone boiling process when red-hot stones transferred heat directly into the mix of water and gruel in the skin-lined cooking pits. This carefully measured fact was unexpected, as archaeologists had long assumed pottery cooking to be more efficient, not less.

Labor costs of cutting and carrying firewood also nearly *doubled*. As nearby brushy firewood became exhausted, males had to find and cut fallen timber from the uplands and haul it to gardening villages. Dried corncobs, fortunately, could be used as substitutes for fuel, as they burned hot and were easily set alight. Cooking with pottery was rather like the modern use of the microwave oven: convenient, but electricity and fossil-fuel-wise, an environmentally expensive nicety. Pottery cooking did eliminate the costly female work of constantly fanning ash away from the bubbling surface of the ancient cooking pit. It also reduced ash-tainted food. A safer and simpler process, it allowed a child to assume cooking duties, thereby freeing up women to pursue other tasks, like gardening.

Pithouses in the Basketmaker II Era circa 400–500 AD

During this era, larger families built larger, more permanent shelters. Men invested more work calories not only in gathering and hauling firewood but also in cutting and carrying the long, heavy poles needed to roof camp shelters or the roofs of deep pithouses.

The Late Basketmaker II's transitional era ended with more sophisticated pithouse construction. Those modifications required unprecedented food calories to dig out a round or oval circle to an average depth of three or four feet, then use that soil in a later stage of construction. The digging cost roughly 350–450 work calories per hour, the nutritional equivalent of about seven consumed corncobs per hour.

Next came the shaping of walls and roof, which rested on six to eight stout, upright poles.[5] These were the main beams, tied together in a square or rectangle to the top beams needed to support the roof. The roof itself consisted of thinner crossties that supported dozens of long, thin poles formed above the ground sidewalls. Thin, flexible branches were then woven in all around to create a woody skeleton for the brush and mud outer wall coatings. The roof received the same treatment. Then came the final coating of mud plaster walls. The mud was mixed from the clayey soil that had been dug out to create the three-to-four-feet-deep subsurface pithouse floor and its interior bench. Most of the primary upright beams were supported on small rock foundations.

When finished, the mudded outer walls and roof had a thermal value of roughly R4 to R6. The clay bench inside—around three feet deep by two feet

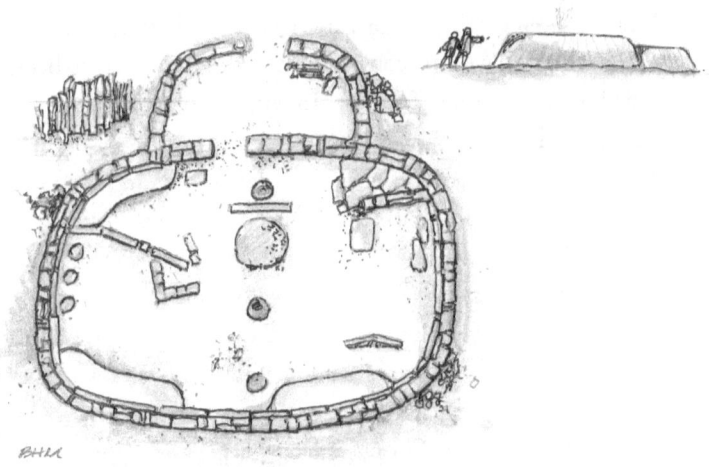

Pithouse: plan view. Drawing by Baker Morrow.

wide—had an even higher R value. Inside, the sunken living space's R-value was higher still and maintained a year-round floor temperature between sixty-two and seventy-two degrees in most middle-elevation settings of 5,500 to 7,000 feet (Stuart 2014). The highest floor temperatures came in mid-fall, while the floors' lowest temperatures came in mid-summer. The summer sleeping floor was ten to twenty degrees cooler than the soil outside. Pure Indigenous genius!

In hamlets above seventy-two hundred feet in elevation, the subsurface floors were often dug out even deeper, to around four feet. Though it took a large pile of wood, branches, green juniper brush, mud, water, and labor to create these snug pithouses, they captured hundreds of thousands of metabolic heat calories from the residents living inside. This architectural style offered some relief from the extreme cold of the San Juan Basin winters. Over the lifespan of such a pithouse (often one to two hundred years or more), millions of human metabolic calories were conserved by the sunken floor and the R-values of the timber, mud, and brush shell.

Most of this period's pit house dwellings were entered through tunnel-like entryways facing south, southeast, or southwest to capture daily sun. The entryways sloped gently downward via a ramp into the interior. The ramp was separated from the living area by draft-reducing interior wing walls and storage bins. Inside, headroom was typically about seven feet. A central mud/stone hearth, two or three grass-lined storage pits, set against mud/wattle wing walls (draft control) and a ventilator shaft to bring in fresh air rounded out the internal features. In winter, cold downdrafts often kept the living spaces smoky.

A very rough estimate of the labor investment suggests that a team of four to five healthy adults could build a pithouse shell in about four weeks, once materials had been gathered. The exposed upper portions of the pithouse wattle and daub roofs and outer walls typically required re-mudding once or twice a year. The support poles from a number of these early AD abandoned pithouses were salvaged and reused in much later house construction—this has made it hard for specialists to assign accurate dates to many sites. Some late Basketmaker/early Pueblo period pithouses also appear to have been ritually burned upon abandonment, during which time the pithouse domes collapsed, sealing the floor features (including remains of hearth and

Early Four Corners community, AD 500–800. Drawing by Baker Morrow.

household goods). Speculation on causes for abandonment ranges from attacks to intentional abandonments with religious overtones. I think it plausible that bunkers of highly flammable dried corncobs, serving as fuel for winter hearths, sometimes caught fire due to sudden winter downdrafts.

The Basketmaker II to Basketmaker III Power Phase

The 200–600s AD displayed all the elements of a culturally and technologically driven power phase, making way for the Basketmaker III era. Cultural changes happened faster. Innovations abounded. Population grew. Human work calorie inputs exploded. The elements of an ecologically based class system emerged, as well as the first hints of religious and political power. The Basketmaker II period had not focused on architectural grandeur, but the raw numbers of pithouses and small farming hamlets established were astonishing in comparison to the San Juan Basin of 200 BC–200 AD. Population continued to rise in the Basketmaker III period, women's gardens complexified, male-tended cornfields expanded, and several new varieties of corn were nurtured—big-eared, drought tolerant, deeply taprooted (planted in sand), blue in color, speckled red kernels, short growing season varieties—140 to 160 days instead of 180 to 190 days—and planted in the higher elevations of seventy-two hundred feet or more.

Pottery became even more refined. It also became increasingly regional in design and an increasingly valuable trade item. Dugout pithouses provided better shelter. New forms of tools were refined for specific tasks.

Semidomesticated turkeys and their eggs were relied upon in several upland districts. Modest streamside irrigation techniques had advanced and population increased, thus the raw number of archaeological sites increased dramatically.

What remains to be measured in the realm of anthropological archaeology, after decades of academic focus on pottery and tools, is human well-being. How long were people's lifespans in the deep pithouse villages? How did that compare to the lifespans of those living in above-ground Unit Pueblo settlements? Who amassed valuable trade goods? Who stored the most food—female gardeners or male corn growers?

A pithouse site yielding shells from the Sea of Cortez is *not* an ordinary site of the late 600s AD, as Chaco Canyon scholar Nancy Akins published in her groundbreaking monograph, *A Biocultural Approach to Human Burials from Chaco Canyon, New Mexico* (Akins 1986). Nor is one sited on a stream bank adjacent to water control features such as small dams, diversion channels, stone and timber dams, and proto-acequias.[6] What is such a site's ecological, technological, and wealth profile as compared to one lacking water and sited at the base of a sand dune blowout five miles away? The Basketmaker III period between 400–700 AD in the San Juan Basin merits much more research.

Ecology, food production, and well-being (including height, weight, longevity, infant mortality, and nutrition) all factor heavily on social organization and cultural values. Societies in which the average age at death is increasing are typically more peaceful, more efficient, calmer, and less stratified than ones in which age at death is declining. I explored this reality in some depth by analyzing trends in height and longevity in the early United States in my book *A Fragile Legacy of Well-Being* (Stuart 2019). When the United States was founded, male heights and longevity had been steadily increasing (female data were simply not available from that time period). American soldiers of the Revolutionary War era were three to four inches taller and lived more than a decade longer than the military recruits sent by English King George III to end the American rebellion. Superior American longevity was a huge colonial benefit.

Societies in which body size and longevity are shrinking are typically plagued by angry disputes between the "doing quite well, thank you" versus the "we are struggling" elements of society. Such societies tend to obsess on wealth, operate wastefully, tolerate a small, rich, secretive authoritarian class, and deny the keys to betterment from wide swaths of their populations.

In contrast, efficiency, transparency, rising longevity, and healthy children are coin of the realm in longer-lived societies. These same factors had begun to shape a coalescing Chacoan world by about 800–900 AD.

CHAPTER 7

FROM BASKETMAKER TO PUEBLOAN

THE AD 700S BEGAN in the throes of the robust but ageing power phase that began in the 500s BC. Family size had grown, pithouses were larger, deeper, and better built, but structural wood was extremely scarce. Turkey feather blankets may have become more common; the birds' carefully layered feathers were bound into a webby matrix of thin yucca fiber cordage and provided a huge, fluffy thermal blanket that helped to stabilize its user's nighttime body temperatures.[1] Shivering costs calories. Those blankets protected infants, pregnant moms-to-be, and the elderly in the depths of snowy winters. Metabolic calories saved by those warm turkey coverlets moderated the centuries-long, late-winter cycle of weight loss. Every ounce of stored body fat helped the Basketmaker peoples through frigid, food-lean, upland winters.

Complexity Increases

The Basketmaker periods span the time from about 500 BC to the mid-700s AD. In most archaeological texts the Basketmaker culture and its people are further divided into early Basketmaker (BMII) and late Basketmaker (BMIII), succeeded immediately by the Pueblo sequence on the basis of its aboveground architecture. Though innovations in architecture were constant, scholars named different cultural periods in ways that often ignored temporally overlapping architectural styles.

The shift from Basketmaker II to Basketmaker III architecture was characterized by larger, deeper pithouses; larger storage areas; thinner and more finely polished, slipped, or decorated pottery; fewer remains of large game; and the presence of distant trade goods. Such goods include shell from the Sea of Cortez, obsidian from the Jemez Caldera ninety miles away, pottery traded north 250 to 350 miles from the Mogollon/Mimbres world in forested southwestern New Mexico, turquoise, amulets made of glassy rock, obsidian blades, and painted wooden sticks, both ceremonial and defensive.

The curved "figure five"-shaped throwing sticks noted in archaeological reports have been assessed as rabbit killing sticks and weapons to fend off lances or arrows. Distant resources, far-flung trade, and fresh ideas of lifeways all flowed into the San Juan Basin during the 500s–700s AD.

Weather-Shaped Architecture and Water-Shaped "Sharing Networks"

The co-existence of several distinct architectural styles of late Basketmaker pithouse structures—one-lobed, two-lobed, lobe and ante room, and variations—all shout growing complexity. Between the 500s and 700s AD, several distinctive architectural styles, variously labeled as Basketmaker II, Basketmaker III, Pueblo I, and early Pueblo II, coexisted in the southwestern arc of the region that would in several hundred years become the Chacoan world's "corn belt." This belt swept southeast from well-watered settlements like Peach Springs at the base of the Chuska Mountains, arced its way east to the Puerco River, went further east to the Rio San Jose, then turned southeast to the small, muddy Rio Puerco.[2] This broad district soon became the first regional source of large cobbed corn available for distant trade.

But progress and growth waxed and waned during the late 700s AD as capricious rainfall patterns tormented the San Juan Basin's corn-growers, bringing drought, then flood. Planting seasons, temperatures, the timing of snow melts, and the first spring rains were all essential factors in planting and harvesting. Thus, any dramatic back and forth in weather patterns—such as April snows after a warm late March—interrupted food storage efforts, diminished regional trade and, in one dramatic episode, ruined entire upland villages.

Tree ring growth charts of the early-to-mid 700s AD indicate that precipitation levels were higher than the San Juan Basin's long-term average. Full storage pits and several decades of rising annual precipitation probably stimulated optimism among farmers in the region. More pocket gardens were planted and hillside corn plot communities grew. Cornfield size increased. In the higher elevations, abundant moisture enhanced piñón nut harvests. In the middle elevations of fifty-five hundred feet, small farmsteads produced larger-than-average corn crops.

In the early 700s AD, the future may have looked bright to small-scale farmers. A number of small clusters of houses appear to have been built on lower elevation lands that had once been over-foraged; after a few years of stable precipitation, some barren plots had again become productive in grass seed gathering.

Unit Pueblos of Pueblo I and II

In the midst of several decades of favorable precipitation in the 600s–700s AD, a new architectural style of aboveground, laid stone and/or wattle and daub dwellings popped up in neat rows near rivulets or small arroyos, just as mushrooms spring up among the dying roots of an aging tree. Formally, these Unit Pueblo homes are known as Prudden Unit Pueblos. Architecturally, they distinguish a transition from Pueblo I to early Pueblo II architecture. These were not dugout pithouses but above ground dwellings. Therefore, they were cheap and fairly easy to build and required fewer thick structural timbers, which had become scarce in the south-central San Juan Basin. The ground-level dirt floors of these two-room houses gave up most of the former sunken-floor pithouse temperature efficiencies. Unit Pueblo style houses and villages housed many "small house" populations of the late Pre-Chacoan era.

These single-family homes had two square or rectangular aboveground rooms connected by an open doorway. The wattle and daub construction of walls needed regular repairs, as thin wood supports were woven in between the uprights (the wattle), then clay-mudded brush and lath walls (the daub). This house form required more frequent upkeep than had the earlier pithouses, but it was inexpensive in both labor and material costs.

These two-room dwellings typically had rather low, six-to-seven-foot

Early two-room suite. Drawing by Baker Morrow.

interior ceilings. Based on analyzed floor detritus, the front room was the living room. This is where people slept, ate, planned the next day, manufactured woven goods (sandals, etc.), and in winter cooked or ground corn. Amenities consisted primarily of mats and a central hearth. The rear room was a storeroom. Storerooms were usually floored with flagstones or compacted clay. Large clay *ollas* (fired clay storage jars) with stone or clay stoppers secured the food from pests.

In the warmer seasons, most activities took place on leveled outside patios: corn grinding (young women), pottery work (women and men), corn husking (everyone), flint knapping (men), etc. As pottery making became more widespread, a home's exterior features often included clay firing trenches, large dugout storage cysts, and a pocket garden of diverse food and medicinal plants. Larger corn plots were typically sited several hundred yards from a cluster of these small, cheaply built settlements during the 700s AD.

Most of these dwellings were sited on gentle hillocks to avoid cold basin floors.

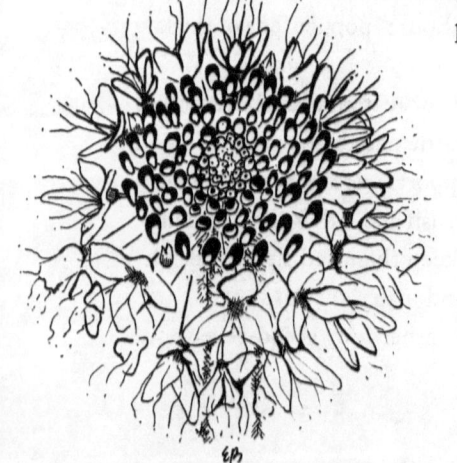

Beeweed sketch by Esther Burton.

The flat timber and adobe/clay roofs were built sturdily enough that they were usable as both work platforms and places to hang drying corncobs. Large rooftop yucca fiber mats also worked well as family members rendered squash meat into sun-dried strips, much like the famous, air-dried "peach leather" staple of Appalachia's European settlers in the 1700s to 1890s.[3]

In the ancient San Juan Basin, long strands of drying corncobs, their husks pulled back, cascaded down from rooftops to just above dog or child height on south-facing house walls. Dried fruits, berries, squash, and corn held up well in the storeroom so long as the mice and rats could be controlled—stone floors, large clay ollas, and stone or clay stoppers all helped. The front room was home; the rear room was the larder; the rafters were the closets. And the external storage pits accommodated food overflow—essential to drought survival. Drought-resistant seed corn varieties were stored separately, as they had high value.

By the mid-700s AD, larger exterior stone-capped storage pits had become common.[4] They may have been community property, as they were in full view of the shallow arcs of homesteads. If raided or looted, the event would have been in full view. Was this a community's stored seed corn for the year to follow? Was it protection against famine? Or had some villages set aside dried corn as a community trade item? Museum-held corncobs need to be tested to identify the locations in which they were grown.

Whatever the specific answers are, large community storage pits were probably reassuring to associated families. Unit Pueblos were built in four or five weeks with cheap, easily obtainable building materials, limiting the need of costly, ever scarcer timber. In contrast, small Basketmaker III pithouses (575–725 AD) needed a month or two for construction and required scarce structural timber to support a domed roof.

TABLE 5. Traditional Dating of Architectural Periods in the San Juan Basin

Basketmaker III	575–725 AD
Pueblo I	700–900 AD
Pueblo II	900–1,100 AD

A Compression of Architectural Styles

The quickly spreading architectural trend to build Unit Pueblos was driven in part because of rapid deforestation. Unit Pueblos required less brush and structural timber. The late 600s to mid-700s was a confusing era. Thousands of trees were consumed, and four distinct, temporally overlapping styles of domestic architecture had emerged: (1) sunken round or oval BMII style pithouses; (2) large, BMIII double-lobed pithouses; (3) two-room Pueblo I (PI) aboveground wattle and daub masonry Unit Pueblo farm houses; and, finally, (4) Pueblo II (PII) wattle and daub above-ground houses, built in compact arcs to create secure farming hamlets.

Just northeast of present-day Gallup, New Mexico, all of these architectural styles were occupied simultaneously, as they were in several other rapidly changing San Juan Basin districts. The older BMII style habitation was often sited on sloping mesas and prominent hills, while the newer BMIII styles were sited on lower hillsides. Both the Pueblo I and II homesteads were sited on small hillocks twenty to forty feet above the dry, cold, and dusty valley floors.

This mix of dated, coeval architectural styles suggests that change and expediency had combined to define a century that defied the neat archaeological/architectural time periods charted in so many textbooks. All four building styles combined living quarters and interior storage areas. All had

Pithouse and Pueblo II—III housing types. Drawing by Baker Morrow.

packed clay floors. The deep BMII pithouses with sunken dirt floors were labor and temperature efficient. The BMIII double-lobed pithouses were larger and more expensive to build, but they housed larger families and could remain in use for several hundred years. The new, two-room masonry late PI and PII Unit Pueblos initially cost fewer work hours and were built with cost-effective stone and clay; however, they had a much shorter use life than the deep Basketmaker pithouses. Over time, subfloor buffering became less common, as many villages had dug out deep kivas and winter houses.

The central San Juan Basin's increasing scarcity of wood and brush must have played a singular role in creating an architectural style that regulated seasonal human body heat. Winters in ground-level dwellings left inhabitants much colder than in early pithouses, and much hotter each summer. Indeed, the collision of coeval architectural styles built in the west-central San Juan Basin area tell us that the pace of change had sped up and increasingly depended on local resources.

Architectural Complexity in the North

During the same time period (670s–870s AD), a similar collision of alleged time-distinct architectural styles appear on the San Juan Basin's northern frontier. Noted scholar Paul F. Reed synthesizes the various Basketmaker periods' styles in the ancient villages in and around the Mesa Verde district of Colorado. At a BMII site coded as "5DL310," a classic early pithouse had been in active use as of the late 670s AD, though it allegedly dated back to the 500s AD. Nearby, Reed cites larger, more complex settlements that were built a bit earlier than they "should" have been (2011).

Thus, we have another example of a district dotted with competing coeval architectural styles that overlap in time and location: the far northern margin of the San Juan Basin with a slightly later architectural style overlapping on the basin's south-central frontier.[5] Imagine the San Juan Basin as a giant geographical billiard table, with the Mesa Verde district as the northern "far rail" and the Red Mesa Valley as the southern "near rail." It is almost as if an ancient pool hustler smoothly stroked a cue ball to the table's far rail. The white cue ball, of course, would hit the far rail, bounce off, then spin back toward the hustler at the near rail. Yes, a cue ball's mass/weight allows it to momentarily store energy when fueled by the energy of a pool cue stroke. Architecture also stores energy, while geographic borders define energy zones.

So, in the San Juan Basin, the frontiers were not just distinct architectural locales; they were energy active ecozones, each creating their own reaction to cultural collision with neighbors unlike themselves. Do such compressions at the geographic frontiers of a culture really influence the trajectory of that society? Just follow the evening news and the dilemma of immigrant families fleeing failed societies because their landscapes had been ruined.

The Wild Climate Shifts of the Mid-700s AD

The late 600s AD had seen a decrease in precipitation in the San Juan Basin. That appears to have pushed some populations into higher, cooler elevations. Throughout the early 700s AD, precipitation was rising. Then in the mid-700s AD, weather patterns became strikingly erratic. Winter snowpacks were trivial, and brutally hot summers were bone dry.

The climatological chaos of the mid-to-late 700s AD reshaped cultural trends, leading to expansions of pocket gardens and increased plantings of large corn plots. In this process of expansion, status distinctions rose dramatically; the obsession with storing enough food for more than a year's needs translated to potential trade commodity and wealth. Storing food was also the only way to create a temporary psychological sense of security. In the lower elevations, nearly all who gardened on dry lands needed not just full storage pits but several regular trading partners who lived in climatologically or ecologically different districts.

Dry-farming valley dwellers in the San Juan Basin needed high-country trading partners to obtain high-grade protein/meat, piñón nuts, and acorn meal. Commodities included dried meat, turkeys, heavy structural timber, and reliable seed corn. Daily needs drew people in starkly different ecological and climatological districts to cooperate with one another. It was *need* and *fear* in the face of recurring food fragility that drove cultural dynamics in the mid-700s AD. Erratic climate had generated an organized upland-lowland trade pattern.

Catastrophe in the Late 700s AD

By the 760s AD, life had suddenly become riskier, thanks to the previous decades' erratic weather. More families were driven to high-elevations to create new settlements in an early Pueblo II style. Villages sprang up on mesa tops and hillsides as far north as Mesa Verde and the forested hills overlooking Colorado's Montezuma Valley. Some of these upland mesa-top settlements consolidated into large villages populated by hundreds of residents, perhaps drawn together for safety in the midst of food shortages.[6] Many of these beautifully built upland settlements included large reservoirs. New upland pithouses were built, shallower than in prior eras because mesa-top soils were only several feet deep above solid stone.

Costly risk management strategies abounded. In addition to the hilltop reservoirs, wood palisades on village margins and nighttime lookouts were built for protection. Evening climbs up cliffside routes to those lookout stations first involved creating hundreds of gouged-out hand and toe holds in the living rock. All of this new infrastructure demanded high protein and extra daily food calories. Still, these risks may have been worth it for a settlement's sense of security amid regional food poverty.

That world's sense of security imploded in the late 700s AD. Sudden grief came in the form of heavy midsummer rains that ruined crops. Mesa-top reservoirs flooded and destroyed well-built villages. Many large villages fragmented into small, desperate groups as hunger combined with violence crushed social cohesion. By the 780s AD, many mesa-top villages were abandoned.

Lessons Learned

Virtually every logical human response to climatic chaos was defeated during the mid-to-late 700s AD. The very large villages that were built for security were hard to sustain. Likewise, very small villages were hard to protect from raids. For many inhabitants in the middle elevations, a cheaper Unit Pueblo architectural style meant they could conserve building materials and food calories—yet they could not resist the instinct to store more dried crops than needed for one winter season. And so food storage capacity expanded by the late 700s AD. These villages also expanded gardens, added outlying corn plots that required more storage bins, and retained the resolute willpower to subsist on low rations, even when full nearby external storage bins were in easy view.

Despite these efforts at saving and storing food, "hunger years" still occurred in a large portion of the San Juan Basin's "small house" population, thanks to erratic precipitation, raids, chronic anxiety, and declining health resulting from corn-based diets.

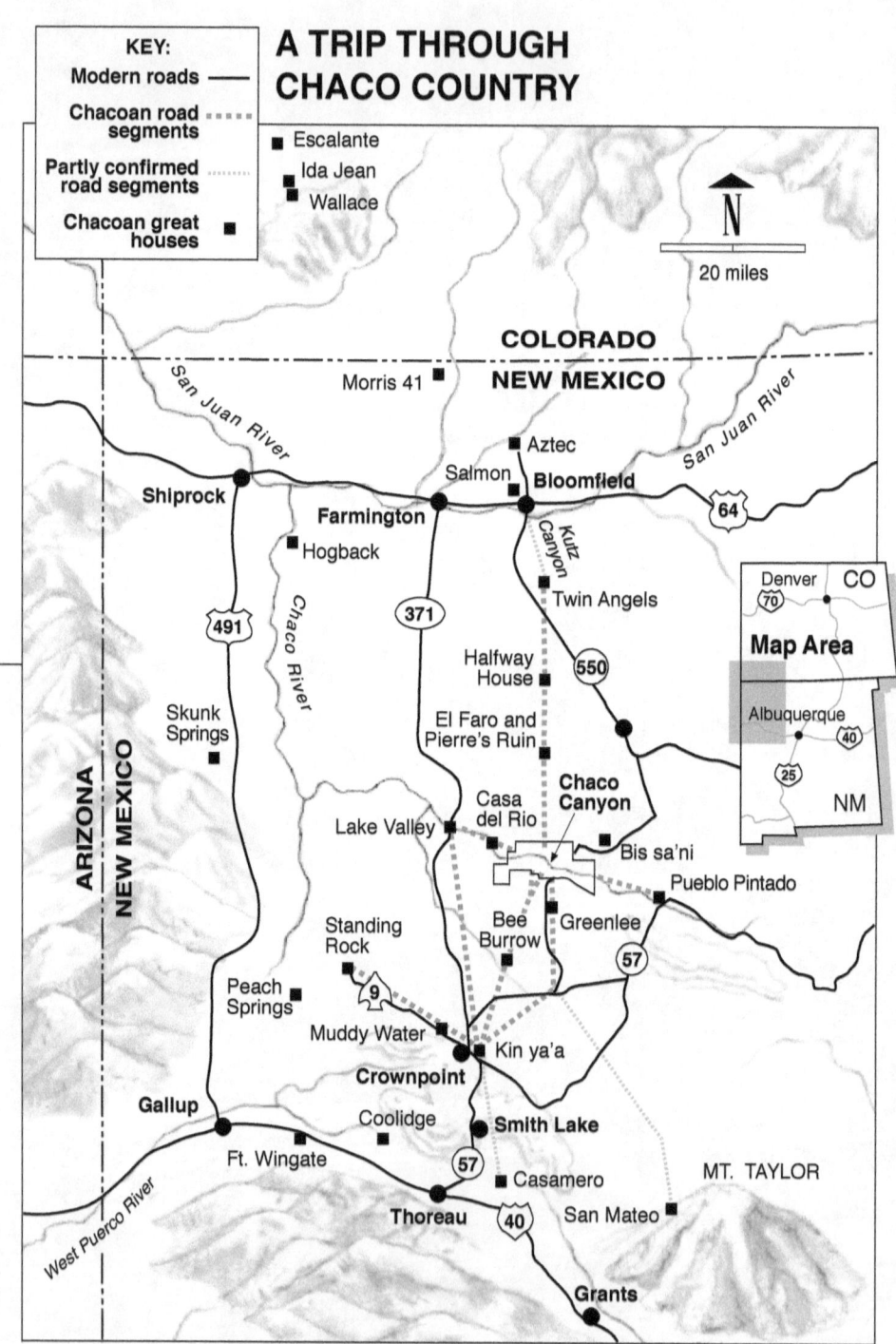

Map 2. A trip through Chaco Country. Map by Carol Cooper Rider.

PART III

SMALL HOUSE FARMERS & GREAT HOUSE ELITES

CHAPTER 8

GEOGRAPHY AND CHACOAN SOCIETY'S EMERGENCE

PROTRACTED SUMMER DOWNPOURS COULD, and probably did, continue to flatten district farmsteads and change local landscapes between the 700s and the 1000s AD. I remember the field surveys south of the Chaco River in 1974 when our crew from the University of New Mexico's Office of Contract Archaeology scampered in and out of a deep arroyo, noting dozens of early pithouse remains that were exposed in the arroyo's water-cut walls. We were focused on the striking cross-sectional views of naturally dissected farmsteads, all 1200–1500 years old. Ancient hearths were clearly visible, as were occasional *manos* and *metates* in place near the dwelling's gray, ashy fire pits. Rain and thunder intensified, and the arroyo bottom channeled a roaring sound down to us. We scrambled up the arroyo's muddy slopes just minutes before millions of gallons of churning mud and water tore house-sized blocks of earth from the arroyo walls. The ground around us shook, the noise so loud we could not communicate for several minutes. When the storm subsided, we retreated to our crew truck, soaked and cold. When we returned the next day, the whole area we had been studying had changed. Five to six feet of passing mud had erased the most exposed ancient farmsteads—others were newly hidden under tons of wet, ashy soil.

Societal Complexities in the 800s AD

Recall that the late 700s AD had been highly unstable, with no single solution or path to security for all. The unpredictable weather had wreaked havoc, and the social rules had changed so fast that fear and confusion were as powerful an enemy as the natural environment had been. From analyzed human bones of that period, we know there was protein deficiency, as were iron deficiency, overwork, and signs of violence, particularly against women (Akins 1986). The "hunger years" had consumed some districts. Life shortened, conflicts rose, and fear likely destroyed confidence in some community relationships.

Such was the status of regional society in the 800s AD, just prior to the dawn of Chaco Canyon society. How scholars assess and define the dynamics of emerging, large-scale changes in an ancient society really matters, as there is no written history. There are few fine-grained clues to a cohesive daily, weekly, or yearly narrative of what life was like in the San Juan Basin during the 800s AD. To compensate, many archaeological texts avoid discussing the social realities and instead analyze changes in pottery styles, architectural changes, and the basics of tools, housing, and farming.

These analyses are important in identifying technical and structural trends, general cultural changes in art, architecture, and technology, but they do not produce a complete picture. Cultures are human adaptive systems. Humans and their behaviors are the *agents* of the tone of a society. And we shall see, through an intricate look at the physical emergence of Chaco Canyon's Great Houses, agricultural systems, and road networks, a picture of its overall societal and cultural infrastructure also starts to emerge. We shall learn that Chaco Canyon society's tone was increasingly created by its elite members, who generally lived free of famine and consumed on rules that *did not apply to their society's majority population*. The tone was multilayered, including social stratification, the presence of an organized religion, and a need to organize outlying districts into one cohesive unit—all of which would fall under the label of Chaco Canyon Great House society.

The Emergence of the Chaco Canyon Great House

Chaco Canyon society can first be seen through the evolution of one of its early Great Houses: Pueblo Bonito. Much will be said in the coming

COPROLITES

Some years ago, scholar Karl Reinhard published an easily readable report on the then-status of coprolite analysis and findings (Reinhard 2000). The analysis of ancient desiccated human feces is usually conducted on material found in and near ancient habitations, gardens, cornfields, and at stopping points near springs, rock shelters, and seasonal camps. True, most coprolites are lost to rabbits, beetles, and other coprophages. An ounce of desiccated feces often carries stunning amounts of information about nutrition. The nature of a breakfast meal is revealed through microscopic and chemical analysis. Even the district where that food, domestic and wild, once grew (nitrogen and carbon analysis) can be identified. Thus, we know that a few ancient coprolites indicate that some meals included corncobs afflicted with blue smut (*huitzilocatchli*). That tells us that the earliest smutty blue corn not only provided higher protein uptake in the diner's body but also suppressed deadly pellagra-like metabolic deficiencies.

From consumed foods, we also can know which rabbit parts were favored at mealtime. Rabbits were often added to the larder as communities worked through their late fall fields with their throwing sticks. Spoiler—the ancient San Juan Basin peoples favored roasted rear rabbit legs. Afterwards, the bones became tools.

Properly analyzed coprolites can tell us the region where the ancient who defecated it grew up (nitrogen or carbon signature). Laboratory analysis can also identify the wild foods eaten. One of Karl Reinhard's most crucial findings is that the early foragers/gardeners of the San Juan Basin had access to, and regularly consumed, a far more diverse diet than did later Chacoan populations of the 800s-1100s AD (2000).

I amplify Reinhard's astute conclusions by applauding the role of ancient women who pioneered their diverse gardens, producing balanced diets during the millennium that ended at roughly AD 1.

chapters about this site and its importance. Between the 500s and 700s AD, several small, oval-shaped Great House prototypes (Una Vida and Peñasco Blanco) were built on a remarkable site that was later occupied by Pueblo Bonito. Pueblo Bonito overlaid an earlier house inhabited by at least one very rich male and his stash of turquoise and valuable Mesoamerican objects. Chaco Canyon's seasonal floods buried the rich man and his treasure under several feet of new clay. There will be more on this burial and others found underneath Pueblo Bonito later in the text. In the late 700s AD, the core of what became Pueblo Bonito was built, including kiva, storerooms, and some small habitation rooms. By the mid-800s AD, several other early Great

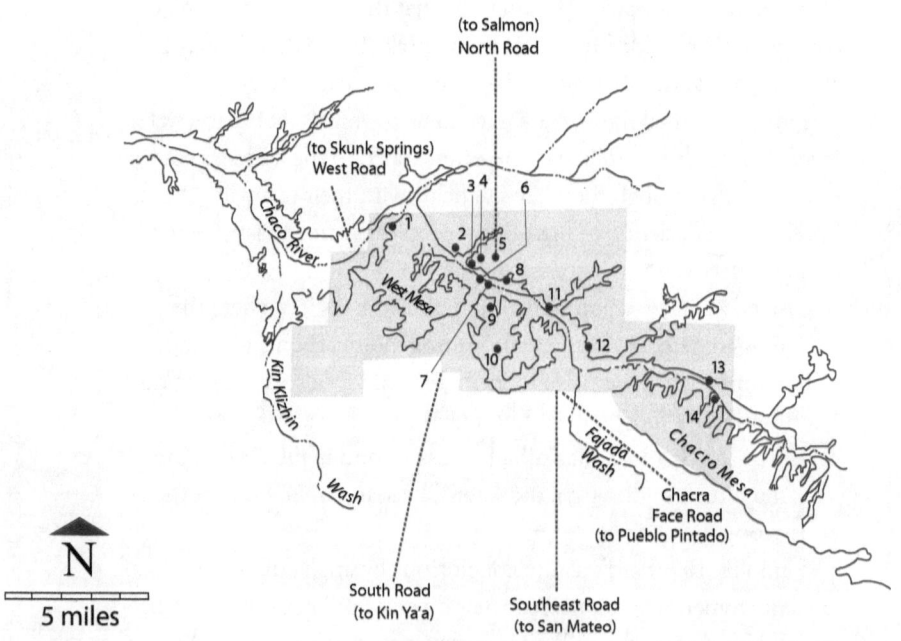

SITES: 1. Peñasco Blanco; 2. Casa Chiquita; 3. Kin Klesto; 4. New Alto; 5. Pueblo Alto; 6. Pueblo del Arroyo; 7. Pueblo Boni 8. Chetro Ketl; 9. Casa Rinconada; 10. Tsin Kletsin; 11. Hungo Pavi; 12. Una Vida; 13. Wijiji; 14. Shabik'eschee

Map 3. The heart of Chaco.

Photo 1. Small house Chacoan architecture. Close-up view (looking west) of BC57 (also known as 29SJ397). This farmstead contained nine masonry rooms, two pit structures *(left and bottom center of house block)*, and two circular kivas *(bottom left and far right)* added about 1120 as social conditions in Chaco Canyon deteriorated. Multiple construction and renovations at farmsteads and many small houses occupied by the peasantry were common. (Courtesy of the National Park Service [NPS], Collection 0001/ VA 722, Negative 119801.)

Photo 2. Views of BC58 *(foreground)* and BC57, two medium-size farmsteads of the 1000s, looking north *(top)* toward the canyon wall and great houses across the Chaco River (denser vegetation near top). Notice the square rooms and circular kiva in the right foreground. NPS archaeologists estimate that it would have taken a skilled person ten months of twelve-hour days to construct it. (Courtesy NPS, Collection 0001/ VA 722, Negative 119806.)

Houses were founded—among them the semicircular Peñasco Blanco and Una Vida, which protected entries to the canyon.

Most Great Houses would come to be built on lands near creeks, washes, and seeps. The hydrology of mesa country also influenced the construction or expansion of an early Great House model that included large, side-by-side masonry rooms in curving rows, and large community *kivas*. Every Great House built over the next four hundred years announced another claim to corn, timber, and human labor as Chaco Canyon's Great House agents muscled thousands of acres to add to its growing political and economic domain.

The Chaco River

The Chaco River pointed like a wobbling arrow flying from north by northeast to southeast in the heart of Chaco Canyon and the western washes of Kin Klizhin and Kin Bineola. The high cliff faces that define Chaco Canyon's north side yielded cascades of falling water during summer monsoons, replenishing a narrow, marshy niche tucked under or against the eroding foot of the Canyon's south-facing cliffs. The lower canyon floor would thunder as masses of roiling water reshaped the river's contours, destroying many small canyon floor farmsteads—just like the scene described at the opening of this chapter.

Those cliffs had been occupied by Basketmaker people or their ancestors since at least the 500s BC, and by Archaic Period foragers even earlier. Narrow, miles-long, food-rich econiches below the cliffs attracted ancient families as early as 2000 BC. Small brush and dirt homesteads endured, generating place-bound lineages of descendants living a few yards from cattail-laden waters.

The narrow ecotone carved out by the Chaco River nourished cattail, a premium food, providing protein, carbohydrates, and valuable fibers, as well as bulrushes, frogs, collared lizards, turtles, and tall, dense wetland grasses. In the less-occupied upper canyon to the northeast, elk, mule deer, and other large game came to forage and water, as they still do to this day. The lower, south-facing cliffs received the most fall sun and could support small garden plots at their bases that prospered thanks to a longer growing season, courtesy of the heat that the warmed up rock radiated from the southern sun.

MY TAKE ON PUEBLO BONITO

Pueblo Bonito's site was probably already a regional legend by the 600s AD. It was an entrepôt of trade in turquoise from the Cerrillos district south of Santa Fe.

The site was magical. It offered fresh water from the narrow rock-meets-water niche at the cliff base just yards from the buried structure's walls. It was a place of cultural exchange with Mesoamerica and was a natural stronghold, the entries to which were likely protected by the Great Houses of Peñasco Blanco and Una Vida.

On the cliff face above the ancient tubrooms, a sky watcher priest had one of the world's best dark sky vantage points with magnificent views of the constellations and their seasonal movements. An ancient sky-watcher's lookout already existed on a mesa pinnacle nearby.

Out of sheer necessity, Pueblo Bonito had slowly but surely learned how to slow water that roared downhill from the upper canyon, limiting the destruction of valley floor farms. In the process of slowing the water, the Bonito engineers for a time managed to maintain a fresh water lagoon adjacent to a large Great House. As time wore on, the extent of water control grew into a wash district of more than five hundred square miles. Hundreds of small up-canyon rivulets, moistened by storms too far away to see, materialized as if by magic. In a semiarid world where water was nearly as precious as corn or seashells, the flow of water from a visually dry sky was staggeringly powerful in the eyes of stressed farmers.

Oddly, those who tended the up-canyon water flows and ritually cleaned the ditches each year are *not* documented in Southwestern archaeological texts.

Photo 3. Aerial view of Pueblo del Arroyo, just east of Kin Kletso, looking toward the north wall of Chaco Canyon (*top*). Chaco River in foreground. Note the well-planned, C-shaped room block and kivas set into square masonry surrounds; these are typical of great houses begun in the 1000s. The courtyard is to the east (*right*). A separate circular, triple-walled structure was built later at the edge of the dry riverbed (*left*). Photo by Charles Lindbergh, 1929. (Recovered by the author from a trash bin in Santa Fe and donated to the Museum of New Mexico, NPS Collection 0001/VA 695, Negative 119778.)

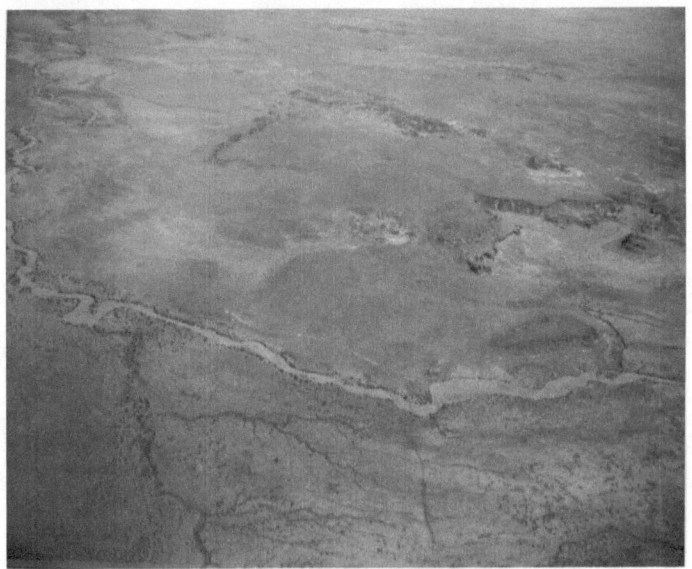

Photo 4. Aerial view of the seasonal Chaco River near the great house called Kin Bineola (whirlwind house). After emerging from higher canyon country such as Canyon del Muerto during the 700s and 800s, the Anasazi farmed these lower, drier elevations where water was available. (Courtesy NPS, Chaco Culture National Historical Park, Collection 0001/VA 693, Negative 119777.)

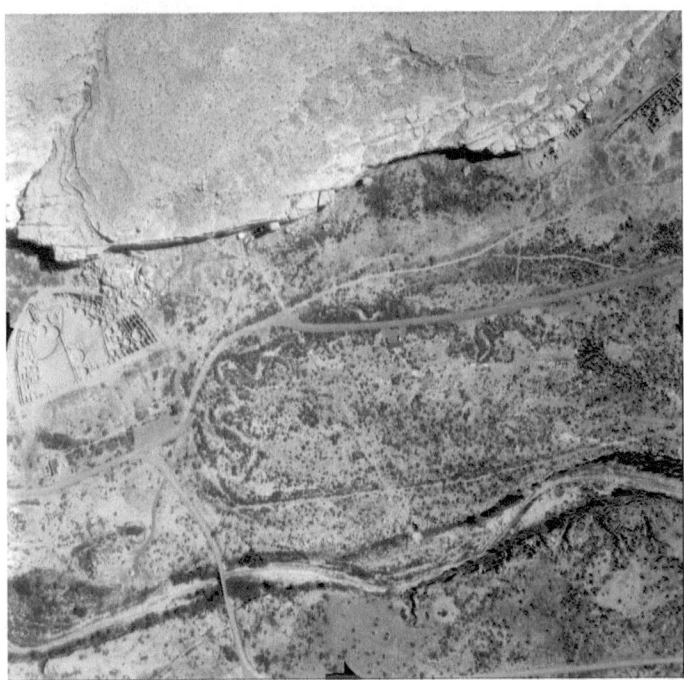

Photo 5. Aerial view of Pueblo Bonito (*left center*) and Chetro Ketl (*upper right*) in central Chaco Canyon. These great houses of the tenth and eleventh centuries were all built on the north side of the seasonal Chaco River (*near bottom*), so that they faced south. The roads (*right center*) are modern. (Courtesy NPS, Collection 0007/015-02, Negative 120215.)

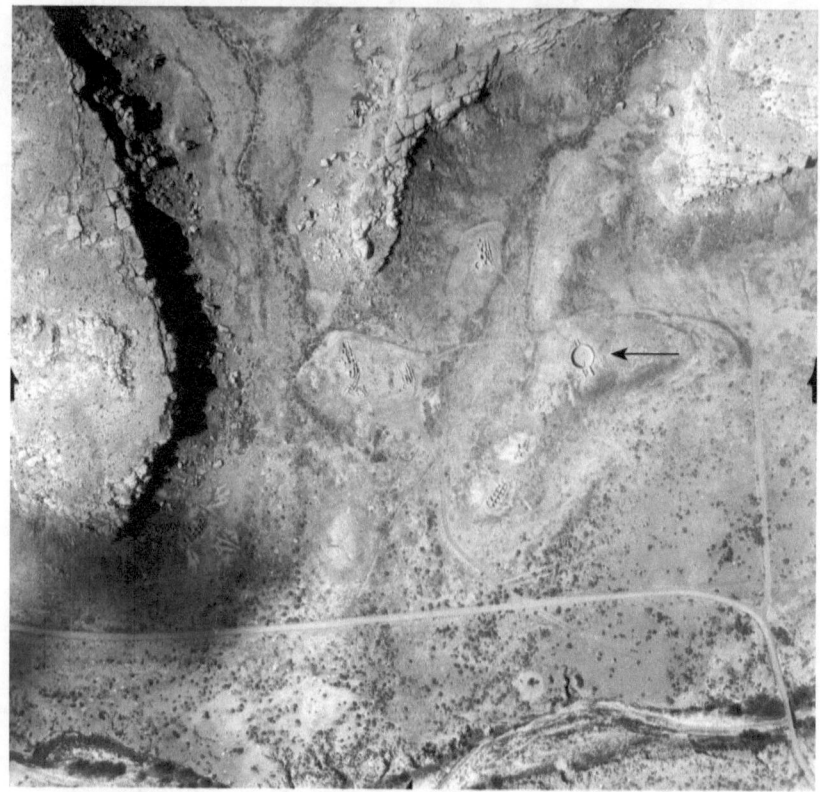

Photo 6. Aerial view of the huge kiva Casa Rinconada (*right center*) and partially excavated nearby farmsteads along Chaco Canyon's south rim, where another arroyo permitted farming. Pueblo Bonito lies just out of view (*north*) at bottom right. Close-ups of several of these farmsteads follow. (Courtesy NPS, Collection 0007/015-02, Negative 120220.)

Dominating this rich ecotone, Pueblo Bonito was the most obvious seat of cultural power for Chaco Canyon. If one envisions the San Juan Basin as a ragged-edged rifle target, Pueblo Bonito is in the bullseye. To this day it attracts the most attention.

A few years ago, a pithouse structure and cooking feature was discovered by the National Park Service. The Park Service needed to expand space for parked maintenance vehicles. The agency brought in heavy equipment to begin excavating an area near the visitor's center by the foot of the

Geography and Chacoan Society's Emergence   141

canyon's north cliff face. Quite by surprise, they uncovered what they judged to be part of an early dwelling structure and a large adjacent cooking pit. Their dozer blade exposed substantial quantities of fire-reddened rock cobbles and thick, compacted layers of ancient charcoaled "brush." This was likely once a small family dwelling that had relied heavily on the above-described ecotone created by the cliff and its narrow stream behind the cooking stones. Botanical remains of bulrushes and other species of cattails were identified. Had the workers extended their digging along the canyon face, they likely would have found remains of other brush shelters and early pithouses.

This site represents one of the rare, naturally specialized ecological

Photo 7. Pueblo Bonito, the jewel of Chaco Canyon's Great Houses. The canyon's south gap and the ancient road to Kin ya'a (*top left*) are behind modern National Park Service buildings, now razed. Note Bonito's multistory standing walls (*left foreground*), curving rear wall typical of the great houses founded during the Pueblo I period, and the courtyard with its immense kivas (*center*). The courtyard was not walled until the 1100s as conditions deteriorated. (Courtesy NPS, Collection 0001/ VA 722, Negative 119797.)

Photo 8. A section of the most ancient rooms at Pueblo Bonito already displaying the trend toward architectural complexity. (Courtesy NPS, Collection 0001/VA 319, Negative 119581.)

niches that were as productive as early gardening. The National Park Service suggested the site to be from the 500s AD. Given the quantity of carbonized brush, it may have produced enough protein-rich plant food to eat and trade without also tending a large corn patch. But it most likely maintained a small pocket garden nearby, as its southern exposure to the sun was ideal.

Chaco Canyon's specialized cliff-base ecotone had likely been inhabited since much earlier times. Special ecological niches are often understudied. Many locales, like the one found during excavation of the NPS truck compound, may have allowed a few families to remain sedentary for much longer periods each year, avoiding the massive amount of work calories spent by walking and moving to successive seasonal camps during the Late Archaic and Early Basketmaker periods. In short, those cliffside dwellers of the late Archaic or pre-pottery Basketmaker period may have been taller and lived longer than later Chaco Canyon inhabitants.

In the 700s AD the Chaco River itself was not the primary source of water to nourish cornfields and gardens. July and August monsoons often sent uncontrollable surges of floodwater downstream at forces that swept away garden plots sited on its banks. In a year when monsoons failed, it did not carry enough water to spread across the valley floor.

In contrast, it was the legacy knowledge of landscape and garden/cornfield techniques from Basketmaker and early Pueblo growers that rendered unlikely portions of the Canyon's sand hills and mesa tops as farm productive. The sandstone mesas in and around Chaco Canyon provided corn growing opportunities on both their sloping table-like tops and sand or crushed rock hills and dunes at the mesa's base. Thin layers of mesa topsoil could be enhanced by cobble or brush and clay "walls" six to eight inches in height, slowing water on the sloping mesa top so that it both sank in and captured soil and plant detritus from upslope. Cobbled gardens were often backed up by rush-bound handfuls of brush/grass. This softened the plant freezing effects of early spring's cold winds and slowed water down enough to sink into the soil, nourishing corn kernels and emerging stalks.[1]

At the base of those mesas, which twisted and turned into rocky corners (*rincones* in Spanish), sandy talus slopes rose from the valley floors. Beneath those sands, derived from eroding sandstone mesas, were typically water permeable layers above even older rockslides, sand, and layers of clay from even more ancient floods. Many of the talus slopes that faced south, southeast, or southwest were farmed. The sandy upper layer of the slopes captured water, the sand then absorbing it and slowing evapo-transpiration. The sandstone's granular rubble provided a foothold for corn plant roots, and the clay layers beneath trapped the moisture before it could reach the valley floor.

Other benefits of Chaco Canyon's physical environs included extensive patches of grasslands and warm canyon niches that sheltered edible plants like sunflowers and edible globemallow. Chaco Canyon was also equidistant from the Chuska Mountains to the southwest and the Sacramento Mountains to the east. Both those mountain ranges provided big game and timber, which Chaco Canyon lacked.

Equally important, Chaco Canyon's location included three major washes (seasonal streams, at times bone dry): the Chaco River, the Escavada Wash, and the Fajada Wash. Each of these wash systems was fed by even smaller rivulets.[2] Best of all, Chaco Canyon lay astride the wandering divide in the vast region's two climatic regimes—the Pacific one that brought winter snows and early spring rains, and the Gulf of Mexico monsoons that brought rains (or drought) in midsummer to fall.

I think of Chaco Canyon like the city of St. Louis in the 1860s to '70s—it was halfway to everywhere and, in the expanding United States, gaining economic and trading power. Yet, in just three centuries, the power Chaco Canyon enjoyed—like that of St. Louis—would begin to fade as the surrounding world of the San Juan Basin aged, fragmented, lost vast numbers of trees, scoured out its foraging patches, and declined in its ability to adapt or generate new efficiencies to restore its overused environment.

The Formation of Chaco Canyon Society

Chaco Canyon society was made possible by a dynamic generated by thousands, if not tens of thousands, of ancient gardening families seeking predictability in the face of erratic precipitation in their districts. At first, small

CHACOAN LANDSCAPE ARCHITECTURE
BY BAKER MORROW

The Chacoans produced not just gardens and later expanding fields, but also other features of an emerging regional form of landscape architecture. The pocket gardens were constructed out of a vernacular tradition and form; so, too, were the Chacoans' extraordinary radial or crescent plazas—seen best, perhaps, at Pueblo Bonito and Chetro Ketl. Features like platform mounds, herraduras, terraces, courtyards, and irrigation works are found in very few other places in the world. New Mexico's first ornamental planting, the great Ponderosa "gnomen," can be found in Pueblo Bonito's Plaza.

Photo 9. Chetro Ketl and Talus Unit (*top left*), tucked against the north canyon wall, were contemporaneous. About half of Chetro Ketl is excavated in this view (looking northeast). The rectangular, multistory great house was built between 1000 and 1115. The great kiva in the courtyard and the circular kivas set within square room blocks (*center*) are features shared with Pueblo Bonito, out of view to the left (west). (Courtesy NPS, Collection 0002/037.001, Negative 25261.)

corn plots, coupled with women's high-diversity pocket gardens, had come to enhance regional nutrition, allowing families to be healthier, live a bit longer, and support more children.

By the 800s AD a decline in dietary vitamins, iron, and protein-rich foods had reduced health and vitality (D. Martin 2000). The all-too-critical sharing networks had begun to fracture as many families exited or abandoned food-sharing obligations that they could not fulfill. As we know from the chaotic decades of the 700s AD, conflict and tension had been on the rise among many early settlements facing food shortages.

Elsewhere, larger BMIII pithouse villages situated in better-watered districts flourished. They coalesced around even larger stone hamlets of prosperous, long-lived lineages. These hamlets, most likely built by community labor, usually included a kiva and, later, subsurface food storage bins.

These prosperous lineages were likely the start of Chaco Canyon's increasingly organized elite class. Some were located on the east face of the Chuska Mountains, miles from Chaco Canyon. In general, communities located on watered land were the most heavily populated. Places like Peach Springs, Skunk Springs, the lower Chaco River, the lower San Juan River, and several of the seasonal washes like Piñabete Arroyo and Captain Tom's Wash carried huge quantities of water in a strong monsoon year. Overflowing stream banks enriched soils.

In the late 800s AD, the increasing small farm populations struggled with food security. When the rains did not come, crop failures stimulated more trade in pottery, obsidian, baskets, and seed corn. Dryland corn gardening families could survive one year of drought by working, trading, and saving enough seed corn for a second year's spring planting. Very few families could survive a three-year dry spell without outside help.

Thus, trading patterns began to shift. The central San Juan Basin's sharing networks had all but crumbled from frequent, widespread crop failures driven by drought. In the place of the old family sharing networks was a new regional paradigm: relationships of intertwined economic and religious factors between humble dry farmers and increasing numbers of prosperous emissaries of the new Chacoan style Great Houses.

The Chaco Phenomenon

The shift from family-to-family sharing networks to Great House trade and protection is now known as the Chaco Phenomenon. It arose under conditions of slightly more predictable precipitation, sophisticated crop technologies, centuries of dominion over exotic, long-distance trade, and the prior centuries of increasing technological efficiencies. Those efficiency gains and the Great House traders' huge geographic reach allowed a small but powerful stratum of San Juan Basin family lineages to replace the grassroots, family-based sharing networks that had existed among thousands of small corn-growing hamlets.

This dynamic is not unlike the one seen in the eighteenth- and nineteenth-century sharecropping world in the southern United States, where the specter of starvation shaped small-scale family decisions and kept sharecroppers

COMPLEXITY THEORY AND THE CHACOAN WORLD

The early Great Houses were successful enough to trigger mimicry, which suddenly opened up new resource markets that spread rapidly over expanding acreages. These geographically expanding Great House resource intakes were modest in many central San Juan Basin districts, but in the aggregate, corn and other foods and firewood flowed into Great Houses, especially in years of abundant late-summer precipitation. This phenomenon was quite like those discussed by John H. Holland and his model of "complex adaptive systems" (Waldrop 1992). When Holland focused on modern adaptive systems during the 1980s, he had also nailed the elements and dynamics of Great House society as it operated between 840–1180 AD.

As Chacoan Great Houses opened up, transactions with small farmers in outlying districts displaced the ancient family-based sharing networks and became the multi-faceted tools of a vast, rapidly complexifying adaptive system. That system consumed and stored all sorts of crucial resources that aggregated socioeconomic power and was probably justified by religious dicta.

Among coveted Great House resources were up-to-date information about landscapes, soils, and crop yields everywhere in and around the San Juan Basin and its outer margins. Information from afar informed the Great Houses and their agents as they planned expansions and alliances. Thus, they also knew which districts were hungry, and which were self-contained. The self-contained were independent; the rest were not.

In contrast, many long-lived sharing networks were starting to lose important information once provided by families steadily being pulled into Chaco Canyon's interaction sphere. As those once-powerful sharing networks shrank and lost contact with faraway relatives, they, too, became more dependent on Great House seasonal information and monsoon forecasts.

The ancient sharing networks were becoming disconnected. That led to reduced information exchanges and a fading regional social structure due to its declining numbers and geographic expanse. Given geographic shrinkage and more tenuous family relationships, entropy and disorder likely rose. Some sharing networks probably considered those kin who had accepted Great House contracts as traitors.

The two food and information distribution systems (sharing networks and Great House food storage) had radically different goals. Corn and other foods held by sharing network partners living in different ecozones circulated diverse foraged and gardened food over wide expanses of the San Juan Basin. The ancient sharing networks had enhanced well-being and protected families in times of famine, drawing them together. In stark contrast, Great House food distribution was treated like bank holdings to enhance its own structure and hegemony. The Great House model overprotected its elite families, reducing their interactions to a significant portion of small farmers.

trapped in relationships with their landowners. It worked for the Chacoan elites for three hundred years. The Chacoan Great House class depended on its members' perceived power to rescue isolated and starving communities and meld their diverse ecosystems into a magically abundant, regional whole. That, of course, would become an expensive illusion as a growing population and longer droughts upset regional dynamics.

CHAPTER 9

THE DYNAMICS OF A SEMIARID EMPIRE

RECALL THAT IT WAS the diverse ecology, central location, potential for a stunningly effective wash district, and known trading paths that, by the 800s AD, destined Chaco Canyon to become the primary seat of a fast-growing world—a dusty basin-like realm that lived on corn, small gardens, scattered piñón-juniper groves, grass seeds, and a burgeoning magico-religious structure that influenced the behavior of thousands of families seeking security. No single individual planned it. No single early overlord reigned absolute over it. Human created power phases are not easily controlled once they are unleashed. The chance to do better or move up is hard to resist as a culture expands, creating new trading niches and occupations, and as basics like food, water, and wood rapidly rise in value.

The San Juan Basin's semiarid ecology imposed many uncontrollable limitations on its human and animal residents. Squeezing a livelihood from two fundamentally stochastic climate regimes (random storm locations and the erratic timing of rainfall and snowpacks) required a cultural mechanism to even out subregional successes and failures and protect the results of success for the most fortunate.

Spotty, irregular precipitation required much more food buffering than could be accomplished by either a single farming family or a small hamlet taking on an ever-increasing number of trading partners. Those partnerships must have involved obligations to share food/seed corn, which could not be guaranteed when both small pocket gardens and larger corn plots failed in a growing season when their trading partners' crops also failed.

In the ecologically and climatologically diverse San Juan Basin, the only way to avoid local- or district-level food disasters was to increase the geographic size of a risk pool and maximize the number of farms and gardens included within it. In the 800s–900s AD, Chaco Canyon elites began to play that singular role by drawing more and more outside districts into their trading fold. Those gardens and farmsteads located near larger seasonal washes were especially desired Great House targets.

Engineered Efficiencies

Memories of the wild precipitational disasters of the mid-700s likely influenced Chaco Canyon's Great House society to devise more sophisticated water control features. The Basin's northern district was better watered, more heavily forested, and supported a more stable groundcover; it was also a region in which it was easier to predict crop potential in early spring. One had only to scan the surrounding mountains to estimate snowpack.

To the northeast of Chaco Canyon, Great House settlements sprang up along the La Plata, Animas, and San Juan Rivers. These communities were both larger and their occupants' diets more diverse than in the central and south-central San Juan Basin. Carrying bundles of corn harvested riverside thirty to forty miles south and downhill to Chaco Canyon cost fewer human work calories per mile than carrying corn sixty miles uphill to the north from the Red Mesa Valley district. In the late 800s AD, improved and widened trails and, later, better-graded roads increased efficiencies. Indeed, when those roads were walked in the 900s AD, the efficiency gains would have exceeded 38 percent.

Many of the factors in play are not often discussed in archaeological texts that are focused on objects. We do not know whether traded/transported corn was typically husked or unhusked before the mid-to-late 800s AD. Cornhusks were valuable as wrappers for food cooked in family hearths. The dried husks were used as cooking and pottery-firing fuel. Unhusked corn with its pollen and tassels intact was highly valued for use in ritual or ceremonial meals.

Virtually no published sources have compared the weight and human transport costs of (1) stripped and dried corn kernels versus (2) dried cobs with kernels intact versus (3) dried but intact ears with kernels, husk, silk, and pollen intact.[1] There is a considerable difference between the weight of dried seed corn kernels versus whole cobs with and without husk, silk, and pollen—three distinct weight factors and the food calories spent in transporting it. As a matter of practicality, each version of transported corn varied in its nutritional, cultural/ceremonial, and immune system value (Heitman and Geib 2015).

Chaco Roads and Economic Efficiency

In 1977 French archaeologist E. Pierre Morenon used respirators to measure oxygen consumption in an experiment he devised. The rate of human oxygen consumption measured both intensity of labor/caloric consumption and body muscle efficiency. Morenon's experiments with walking the old Chacoan road segments indicated that using thousand-year-old, deteriorated roads was 38 percent more calorically efficient than walking the same routes a few yards outside of the smoothed Chacoan roads (Kantner 2023). That was a huge efficiency in transport costs!

Chaco Canyon society's first major building phase during the mid-800s AD appeared to have marshaled enough regional labor to improve and widen ancient trails as well as build an elaborate Great House like Pueblo Bonito, surrounded by smaller, older farmsteads. Smoother, more easily followed trade routes had injected a major new efficiency factor into the San Juan Basin's central and southern districts. Smooth paths meant faster walking, fewer baskets of goods lost to falls, fewer twisted ankles, and reduced transport time. The emergence of an increasingly sophisticated road system in the central part of the Chacoan world contributed to trade and market growth, due to an expensive human cultural intervention.[2] Why was this so significant? And why did the Chacoan road system expand first in the west-central and southern San Juan Basin? One answer is that during the mid-to-late 800s AD, the central and southern San Juan Basin was producing more corn than other sectors of the basin. Exportable quantities of corn, husks, and cobs would have been produced in the larger corn plots, *not* in the small, woman-tended pocket gardens.

Corn Growing Expansions

By the late 800s AD, corn farming expanded into cornfields of one acre some distance away from a cluster of late PI/early PII family farmhouses. These plots were largely tended by male labor on land owned by their mothers, sisters, wives, or aunts. If the rains came abundantly from an ample July/August monsoon, huge quantities of corn might well be produced, giving a small farming hamlet both stored corn and trade/market commodities.

Planting large cornfields was one way of facing up to a statistical climatological reality of failed rains. Another reason for the "large field" planting strategy deals with reemerging male status, travel, local trade, and access to distant trading partners. In short, the large cornfields represented a power strategy. Meanwhile, women's carefully tended gardens remained focused on efficiency and constancy.

In retrospect, the efficiencies and sense of security gained in large part by women's gardening up to the 700s AD had been dented by both population growth and the troublesome climatic events of the later 700s AD. In the modern world, societies tend to see "progress" as the signals of growth and power. Yet once in a power phase, it is very hard to reverse course. Think of a coal-fired train as the fireman frantically shovels coal to create more steam. The train puffs off excess steam and coal ash as it moves forward, "gathering steam." On a level track, speed and coal shoveling soon stabilize. The train's great weight begins to pull and, like an enormous atl-atl dart, it becomes a segmented projectile. All is calm and well until a long, uphill grade is encountered. There is no alternative to adding more coal, shoveling faster, and often trading brief shifts with a second shoveler to maintain momentum. If the fire and pressure in the boiler is not increased, the distance already gained will be lost and the train will roll backward, driven by its own weight. In such a case, most of the fuel energy used to mount the hill will be wasted if gravity pulls the train backward to its starting point. Starting the train and stopping it wastes more energy than when it chugs along the flat stretches of track at an even speed.

The Food Train

In the San Juan Basin of the mid-800s AD, it appears that large-cobbed corn became the train, and the regional dietary food consumption of the growing population was the caloric slope that generated backslide. Like a train meeting a hill, the metabolic machine of the human body can also run out of steam. In the Chacoan world to come, corn was coal, and its calories heated human bodies, whose labor served as portable wealth/power. The 800s AD were not merely a definable archaeological subperiod in the birth of Chacoan society but also a time when demographic hill climbing reared its ugly head as food supplies failed to fully match the needs of the growing population.[3] That dynamic likely played a role in the shift from family-based trade partners/sharing networks to Great House agents during the mid-to-late 800s AD.

Though the San Juan Basin's total corn production had generally increased between the late 700s to the late 800s AD, its availability remained both unstable year to year and place to place due to erratic monsoons and remarkably spotty rains. This spottiness was the bugbear of most gardeners/farmers in the San Juan Basin.

Dynamics of a New Power Phase

Increasing farm-labor inputs exploded during the 800s–900s AD. Harvesting corn that exceeded family needs had fueled regional trade. New cornfields had to be prepared, and carrying water to larger, more distant cornfields required endless baskets of water. Ever-scarcer firewood also led to notable reduction in groundcover (fuel, cordage, grass lining for storage cysts, basket grasses, etc.), which made life especially hard for scattered, small-scale gardeners/farmers in the impoverished, dry-farming districts situated along the southern approaches to Chaco Canyon.

Among the settlements of what is now Fort Wingate and the sandy district just west of the Puerco River, the "train" of corn farming activity had stalled on the uphill obstacle of insufficient labor, too few dietary work calories, stingy ground cover, and too little rain. Sufficient dietary needs could not be met among clusters of dryland farmers by the late 800s AD. The third and fourth generations of those who had gardened successfully in the 780s to 830s AD were, in health and in well-being, steadily falling behind those who gardened or farmed near reliable ground water, washes, or creeks that overflowed their banks in most years.

In short, the power phase continued to gather steam in the Chuska Valley, Chaco Canyon, and Red Mesa Valley districts but faltered elsewhere. In order for the cultural and economic train driven by small farmers in the 400s–700s AD to make its second hill, it needed to detach its last few cars, or find a more powerful driving engine. Those "last few cars" were primarily the inhabitants of small, dry-farming communities in the driest lands of the southern San Juan Basin. Those farm hamlets likely suffered and grieved from erratic monsoons, felt frightened, and struggled to feel as though they "belonged" to regional society.

Corn and Women's Health in the Late 800s AD

Sadly, young women bore the lion's share of grief with regards to daily life. The grief that swirled around them came in many forms. First, women were overworked. Corn grinding on bent knees was grueling. In fact, it was so intense that it consumed three hundred to four hundred calories per hour. It also slowly reshaped young women's pelvic girdles. A few years of three- or four-hour sessions of corn grinding would flatten the bones of the pelvic girdle from a wide oval to a flat, narrow birth canal. Many young women undoubtedly died in prepartum agony, their infants' heads too large to pass through their work-shaped pelvises (D. Martin 2000).

Female well-being in the 800s AD was in decline throughout much of the San Juan Basin. From skeletal analyses, women in farming/gardening communities suffered greater percentages of substandard diet than did men: far too few food calories, far too little protein, and too few crucial dietary minerals, especially iron.

By the mid-800s AD, there was also an increase in depressed skull fractures and broken arm bones in women, likely due to the actions of their men folk. Debra Martin (2000) notes markedly increasing evidence of male abuse at this time. In addition, Nancy Akins's 1986 monograph cites convincing evidence that women living in the small house farmsteads within sight of Pueblo Bonito suffered high rates of physical abuse.[4]

According to Martin, these abuses disproportionately impacted females and had risen notably by about 800–850 AD (2000). The contrasts in health between men and women also continued to increase. In some districts,

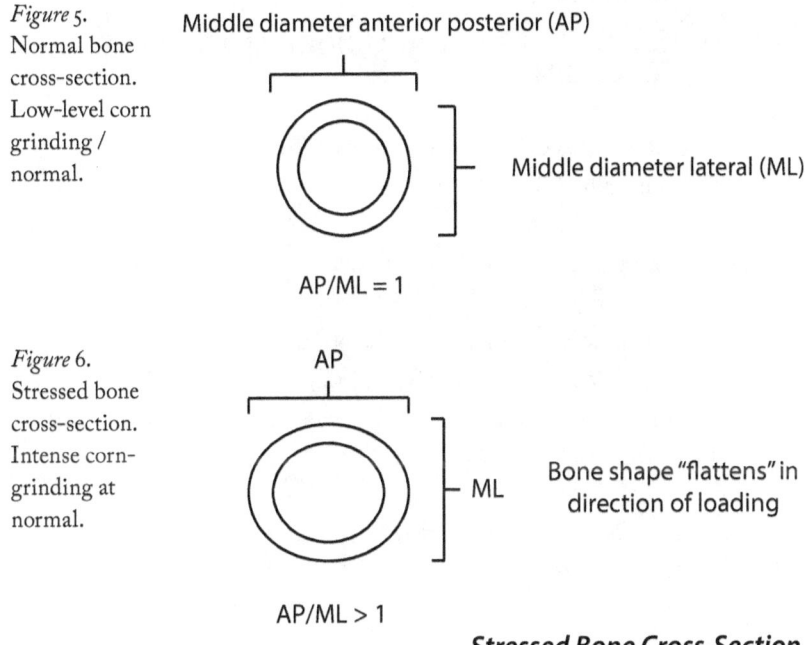

Figure 5. Normal bone cross-section. Low-level corn grinding / normal.

Figure 6. Stressed bone cross-section. Intense corn-grinding at normal.

Martin asserts, women lived a mean average of 21.6 years, and men about 24.4 years. Weakened by poor diet and intense labor, young women in the San Juan Basin at this time did not likely menstruate till about the age of seventeen. With an average lifespan of 21.6 years, this probably limited many women to only two pregnancies in their lifetime. Considering their poor nourishment, they might have produced children with congenital defects and weak immune systems. Breast milk produced by a poorly fed woman can't do the job of fully strengthening a child's body and immune system.

Corn Diet

A diet of 65 to 70 percent corn did not provide enough vitamins and minerals to support women's health. Such a corn-heavy diet enhances the side effects of iron deficiency anemia and pushes the human body toward a dangerous metabolic state. Corn provided ample calories and a brief bodily satisfaction as its sugar metabolized. But the sense of fullness faded quickly. For

> **FEMALE DAILY DIETARY REQUIREMENTS IN THE SAN JUAN BASIN**
>
> 1. The recommended daily allowance of iron for women is 18 milligrams daily. A lactating woman needs 36 mg daily or 3.5 oz of amaranth flour.
> 2. The RDA of iron is 15 mg daily for children and adult males, or 2.8 oz of amaranth flour.
> 3. One ounce of meat and 1/2 ounce of piñón nuts added to cultigens still leaves an **iron deficit** of 10 percent. Without piñón nuts, meat, or amaranth seeds/leaves, the corn-based diet is 40 percent below what is needed.
>
> Source: Debra Martin, *Women and Men in the Prehispanic Southwest* (2000)

millennia, inhabitants of the San Juan Basin had chewed or sucked on fresh grass stems as they went about their daily routines. The quantity of sugar in a grass stem was modest. Dozens of grass stems sucked per day, however, did reduce hunger pangs, though this also contributed to diminishing groundcover and its protein-rich grass seeds.

This trend toward more and larger corn plots modified the food strategies of both traditional gardeners and large-plot corn-cropping families. Had women's highly productive, highly diverse gardens continued to be expanded and the rains become even a bit more predictable, the physiological well-being of ordinary families might have improved. Instead, well-being among the majority of ordinary families declined as the Chacoan Great House elites and their subculture grew in wealth and size.

CHAPTER 10

THE GREAT HOUSE ERA, 875-1175 AD

CHACO CANYON'S SPECIAL STATUS was evidenced by the early star or sun observatory that straddled a nearby incline. Small-cobbed corn had been grown in the canyon since 2000 BC, and several layers of early shelters and pit rooms all rested quietly beneath Pueblo Bonito. Some of the region's early and wealthy traders had been buried with their bounty under what archaeologists named Rooms 32 and 33 of Pueblo Bonito. Three of the earliest burials were radio-carbon dated to 676–894 AD (Skeleton 12), 690–944 AD (Skeleton 13), and 690–940 (Skeleton 14) as reported in Kerriann Marden's chapter "The House of Our Ancestors" in the anthology *Chaco Revisited* (Marden 2015).

In the 600–700s AD, Pueblo Bonito as archaeologists think of it did not yet exist, but the wealth discovered under its floors offers a clue to the vast geographical extent of Chaco Canyon's early trade. The burials in Rooms 32 and 33 were found under the primary floor (Marden 2015). Among the burials were more than twenty-five thousand pieces of turquoise and hundreds of Mesoamerican ceremonial objects like drums, bracelets, and conch shell trumpets (Marden 2015). Turquoise was very valuable throughout Mexico, Central America, and the Southwest, as its blue tones were associated with water and several Aztec gods. Other valuable Mesoamerican trade items found consisted of a pyrite mirror, conch shell amulets, exotic pottery, and quantities of shell beads later called *heishe* by seventeenth-century Pueblo peoples.[1] The precise positions and burials dates of these skeletons have been muddled by outdated laboratory techniques.

Recently, a DNA study of female burials found in Room 32 has energized the idea that a genetic lineage of women owned and ruled Pueblo Bonito. That a female lineage held such property bolsters my assumption that men owned the portable wealth while women owned real estate: houses, gardens, and farms. We do not, however, know who owned the water that flowed past Pueblo Bonito's ancient subterranean tubrooms and storage rooms. The question of who "owned" the water remains unanswered.

The turquoise found with the Bonito male burials may have been chemically analyzed in order to discover where it originated, but I have so far found no specific reference of it. Much of it came from deposits near Madrid, New Mexico. The shell most likely came from the Pacific shores of Baja California or the Sea of Cortez in Sonora, Mexico.

Great House Health Benefits

Analysis of these Great House rooms shows us the immense wealth the elites had accumulated. Forensic analysis tells us that their physical and health profiles were equally impressive. Chacoan elites were, in comparison to their small house neighbors, several inches taller in height. They lived longer and enjoyed a higher level of gender equality in diet, health, and power than members of the small-house farm population scattered along the Chaco River's floodplain. The elites also wore better-woven sandals than the region's outlying Unit Pueblo gardeners, enjoyed more meat and piñón nuts in their diet, and suffered much lower rates of female physical abuse.

The early Great House women had turquoise jewelry and likely owned residential multistory suites in Pueblo Bonito. Ample private domestic space also meant lower contact exposure to communicable diseases. These Great House women were mostly hidden from public view. This may have limited these elite women's awareness of the miserable life conditions suffered by small house citizens within sight of their upper-roof terraces at Pueblo Bonito.

Pueblo Bonito's male priests may have already obtained cacao from trade networks reaching south into Mexico by 750–800 AD or earlier.[2] The import of knowledge and the medicinal secrets of steaming cacao in cylinder jars from central Mexico was a clear advantage—cacao helped keep the Sun priests' pupils clear and relatively cataract free.[3]

The Exceptional Corn Pollen

Great House elites may have eaten corn pollen, one of the most highly nutritional foods available. It is high in protein, omega 3 fatty acids, iron, and magnesium (oddly, corn kernels lack all of these essentials). Specifically, corn pollen has a protein content of 15–18 percent. That protein is primarily lysine, which the human body needs.

Excavators at Pueblo Alto have long noted the surviving quantity of corn pollen in Alto's soils. In addition to elites' meals, it was also served as a sacred food during their ritual gatherings. Corn pollen had become sacred, iconic, and a version of specie and portable wealth in the San Juan Basin. It was also a highly valuable trade item, especially as commerce and Great House construction accelerated in the late 800s AD.

Another Great House Power Phase Gathers

By the mid-800s AD the advantages of Great House culture was established, and a power phase for control over the region's corn economy was brewing. Meanwhile, small house villages were doing what they could to be self-sufficient. When full, their rear storage rooms and large community storage pits could save gardening and corn planting families caught in the grasp of a one or two-year local drought.

But by the mid-to-late 800s AD, the volume of household and village storeroom capacity among small house corn farmers appears to have stalled in the farm regions surrounding Pueblo Bonito. Storage space topped out at about 40 percent of living space and was never after systematically enlarged in most of the scattered dry-farming hamlets after the mid-800s AD. Regionally, small house self-sufficiency was on the decline.

It must have been obvious to the well-fed Chaco Canyon sky watchers, priests, and Pueblo Bonito's lineage of female land-owning elites that offering food security and seed corn to those whose crops had failed in the hinterlands would give their canyon society enhanced status, trade, and religious power. As rescuers of the poor and hungry, they may even have imagined themselves as do-gooders, or utterly essential in preventing their surrounding world from foundering. Whatever their motivation, a far more complex regional society in which Great House elites began to take active control of regional trade was rapidly forming during the mid-to-late 800s AD.

Roads to Great House Expansion

Around the mid-800s to 900s AD, there were notable architectural signals of increasing Great House power: the construction of new, nearby multistory Great Houses where storeroom volume vastly outpaced residential space; religious construction (huge paired kivas in which elites had their own priests, secrets, and storage areas); and the emergence of a formalized road system.[4] It took huge quantities of labor to create smooth road segments, filling in low spots and shaving down high spots in rocky desert floors. By the mid-900s AD, several major roads led to Chaco Canyon.

In stark contrast, the small house hamlets outside of Chaco Canyon continued to lose economic power. Village populations increased and corn plots expanded, while natural landscapes were systematically overharvested for food, fuel, and fiber. In the regional scramble for structural wood, poor families had to trudge great distances to acquire their own medium-sized roof beams. Even the Chaco Canyon's Great Houses had to import virtually all of their construction timber from either the Chuska Mountains or the Zuni and Gallina upland forests. As a consequence of scarcity, long-abandoned dwellings were typically stripped of useful wood and reused in new construction.

As the south-central San Juan Basin's natural resources diminished, the Great House elites were motivated to expand their reach. Pushing outward was essential. The incorporation of new, more distant and thinly populated districts offered the possibilities of more trade in corn, better pockets of groundcover, and new clay sources for pottery.

Over the next few centuries, pulsing geographical expansions of Chacoan style architecture would dominate the region. Smallish outpost Great Houses were built quite far from the Chaco Canyon core.[5] This type of territorial expansion was the only sensible solution to compensate for land overuse and destruction of groundcover. Extracting resources from less-populated districts of the San Juan Basin focused on expanding the water flow of small washes, and patches of slightly better groundcover supported species like rabbits, turkeys, prairie dogs, and occasional larger game.

These outlying districts also provided cheap, local labor for the outpost Great Houses' construction teams. Ten to twenty local men, guided by a Great House engineering team, could connect small washes, reshape field slopes, and direct wash waters to corn plots. The massed labor and engineering techniques enhanced local agricultural potential, and added another corn-driven outpost to Chaco Canyon's increasing territorial grip.

CHACO CANYON DESIGNATIONS

Some scholars define the outer ring of Chacoan style communities as the "Chacoan Halo" (Doyel 1984). That halo likely encompassed boundaries that shifted during Chaco Canyon's history.

To simplify commonly published labels, "Downtown Chaco" means the cluster of Great Houses on or near the Chaco River and its junctures at Escavada and Kin Klizhin washes (Vivian 2015). The Chaco River is now a deeply entrenched, seasonal wash between the Great Houses named Hungo Pavi and the oval-shaped Peñasco Blanco. In contrast, the "Chaco Core" is based on a much larger geographical perspective than Chaco Canyon. The Chaco Core typically refers to the fifteen Great Houses spread along a series of washes from Pueblo Pintado on the east to Casa del Río and Kin Klizhen (meaning "dark house," from the dark shade of its sandstone walls) to the west. This is the main area where multiple seasonal washes collided beneath dramatic cliffs and broken mesas. These washes, east to west, are named Kin Bineola, Kin Klizhin, the Escavada, Cly's Canyon, Mockingbird Canyon, Gallo Canyon, and Fajada.

At the far east end of this core district is the late Great House site of Pueblo Pintado, built about 1050 AD and never fully lived in. It may have been intended as the core of a new Great House community, one that offered the chance to harvest extensive wild grasses and claim land closer to the reclusive and bellicose high-elevation populations that inhabited and controlled the wooded Gallina uplands to the northeast of Pintado (Vivian 2015).

Photo 10. Pueblo Pintado, about thirteen miles southeast of Wijiji, was the first Ancestral Puebloan great house that Lieutenant Simpson encountered in 1849. Probably built as one planned project in 1060–1061, it contains about sixty immense ground-floor rooms and several kivas. A Chacoan road was built to it at about the same time. I view it as a public works project akin to CCC and WPA projects initiated by the US government in the 1930s to absorb idle labor and shore up a failing economy. Those same conditions prevailed in the Red Mesa Valley farming district south of Chaco Canyon at about 1050 AD. (Courtesy NPS, Collection 0002/037.001, Negative 30599.)

Agriculture in the Chaco Core

A recent re-evaluation of the Chaco Core's agricultural potential has outdated earlier assumptions that the core Great House district had a rather low agricultural promise. Not so, according to Vivian and Watson. In a quite clearly argued chapter, they offer a detailed analysis of both ancient corn plot locations and more recent successful Navajo plantings spanning the 1890s to 1950s (Vivian and Watson 2015). The net result of Vivian and Watson's landscape analysis indicates that agricultural potential in the Chacoan core was much higher than most published sources have assumed. That enhanced assessment is quite important, as it points out the benefits of broken terrain, seasonally wet washes, pockets of sandy soils underlaid by clay deposits that retain moisture, and small but agriculturally productive mesa tops and talus slope tongues. These planting strategies all combined between 780–1130 AD to make a case that Chacoan Core Great Houses and their associated small house communities could grow enough corn in most years to be relatively food secure. This reality is in sharp contrast to the small, isolated gardening and farming communities in the south-central San Juan Basin, where famine was more frequent. In fact, the Chaco Core was food secure enough to expand Great House storage capacity by the mid-to-late 800s AD, a time period when archaeologists have identified enough changes in architecture and artifacts to distinguish the PII period from PI.

Competing Subcultures

Those changes were advanced by a mixture of population growth, hardship, evolving human work roles, and enhanced understanding of water flows and controls. By the mid-800s AD sociodynamic realities in the San Juan Basin had become more complex than they had been in the 500s–600s AD. Several culturally and linguistically distinct daily worlds had begun to collide, including the ancient Zuni district, the Mesa Verde district, the Red Mesa Valley and Mount Taylor districts, the San Juan River and Animas River communities, and the Chuska Valley.

Due to a shorter growing season, Mesa Verde and other upland districts produced smaller cobbed corn than that grown in the central and southern San Juan Basin. That meant less bulk and fewer calories per human labor calorie invested. But deer, turkey, piñón nuts, and acorn meal all contributed to their nutritional security. At Mesa Verde, corn never became king.

Over the centuries, Mesa Verde retained its high ecological diversity even as population rose. A wide variety of foods sources is powerful in small-scale horticultural societies. At Mesa Verde, this was due to multiple sources of forageable grasslands and forests, mesas rich in large and small game, available firewood, pottery clays and, on its northern approaches, a vast region characterized by high ecological diversity and lower population density.

Thus, the tone of Mesa Verde life in the 400s to 800s AD was solid, stable, and economically self-sufficient.

This self-sufficiency was of special benefit to the women who lived in Mesa Verde. According to Debra Martin, an adult female needed approximately two thousand calories and forty-five grams of protein per day. Those dietary needs, she reports, could be met by a daily diet of six cups of cornmeal, two tortillas, two cups each of beans and squash, one cup of amaranth seed, one ounce of turkey meat, and one-half ounce of pine nuts (2000). I suggest that Dr. Martin's estimate of female dietary needs be reduced by about 9 to 10 percent to approach an actual ancient Mesa Verde females' needs, on account of smaller body size and unusually efficient metabolism.

With this reduction, we arrive at a diet of 1,800 calories a day, with 39.5 grams of protein from 4.3 cups of corn, one and a quarter tortillas, 1.6 cups each of beans and squash, 7.2 ounces of amaranth, 0.9 of an ounce of turkey meat, and 0.25 of an ounce of pine nuts. Further, we can assume this met the dietary needs of an active adult female weighing 98–106 pounds in the Mesa Verde district of the 400s–800s AD. In recent years, medical data indicate that surviving descendants of early Puebloans are 6 to 8 percent more metabolically efficient than the average modern European descendant now living in the United States. In eras of famine, the most metabolically efficient humans out-survive the least efficient.

Given the relative ecological advantages enjoyed by the Mesa Verdeans until the mid-800s AD, there was no natural push for rapid population growth, nor did they require a massive geographic expansion to provide for a starving underclass.[6] Mesa Verde's only liabilities involved shorter growing seasons, smaller cobbed corn, and potential attacks from more northern plains people. The district's climate was heavily influenced by the El Niño and La Niña weather systems and precipitation included both snow and rain. Several early hillside reservoir structures have been located in that region to capture mountain rains and snow melt.

Shabik'eshchee Village

While Mesa Verde enjoyed its diverse foods and relative stability, Chacoans were dealing with a serious loss of ecological diversity. The need for

MESA VERDE

Scholars have long noted Chaco Canyon's ties to Mesa Verdeans, who had migrated south and were living in a portion of Pueblo Bonito by the mid-800s AD, increasing family complexity there (Ware 2014). The Mesa Verde district provided pottery, large game meat, piñón nuts, and corn to Chaco Canyon. It was also a source of ancient lineages of genuine Basketmaker cultural founders.

A Mesa Verde habitation at Pueblo Bonito may have been integral to Pueblo Bonito's cultural, religious, and ancient, first-use rights of land ownership. Both Mesa Verdeans and Chaco Canyon lineages clearly identified with early Basketmaker culture. Those common roots were expressed in bifurcated basket forms, particular farming techniques, sandal styles, religion, and language. A very lucid account of these presumed connections have been published by Edward Jolie and Laurie Webster (Jolie 2015).

ricegrasses—for Great House interior wall material, beds, storing corn, and dietary protein from its seeds—had likely skyrocketed by the late 800s AD as the lower west end of Chaco Canyon became more heavily populated and Great Houses proliferated. Grasslands in and around the Chaco Canyon Core continued to diminish from over-collecting.

An ancient Basketmaker village called Shabik'eshchee sited atop grassy Chacra Mesa appears to have remained a protected ecological reserve throughout the Chacoan era.[7] Tentatively dated to the 500s–600s AD, this village may have been ancestral to one of the subsequent Great House lineages. Nearly one hundred pit dwellings and many garden grids have since been recorded. Its later use in the 700s–800s AD may have included much-needed ricegrass foraging.

Mesa-top farms were stable havens and very typical of the Mesa Verde area sixty miles to the north. Was Shabik'eshchee village a genetic and cultural factor in Mesa Verde's obvious cultural influence on Chaco Canyon's

Great House world? Did the village, little studied to this day, represent a cultural throwback in Chaco Canyon? Its architecture is typical of the 300s–600s AD. It was never renovated into the high Pueblo Bonito style—multistory, with carefully stacked stone—rather, it seems to have been an important founder's lineage site in a diverse, grassy environment where several washes collided below.

As University of New Mexico scholar Wirt H. Wills has observed, Shabik'eshchee occupied a prime location for agriculture. It may also have been valuable for its surrounding mesa-top grasslands. Those grasslands would have provided jack and cottontail rabbits, Indian ricegrass, patches of edible globemallow, chenopod greens, and amaranth. In its preserved state, Shabik'eshchee may still protect hints of an early Great House society in which women's gardening and their foraged food resources characterized a golden age of diverse pocket gardens and more adequate diet. Did Shabik'eshchee also focus on trade goods from afar, as did Pueblo Bonito and Chetro Ketl in the Chaco Canyon core? What cultural values did it represent to the Great House elites? Why did it become a protected reserve?[8]

The eastern or "upper" end of Chaco Canyon between Shabik'eshchee and Pueblo Bonito is appealing. It radiates a sense of calm. There are still soggy cliff-base patches full of bulrushes, occasional tiny frogs, grass lizards, deep stands of various grasses and desert flowers. Both jack- and cottontail rabbits abound in the zones where thick grasses meet drier soils and the pink or lemon tones of edible globemallow near today's dry gravel road. One afternoon some years ago, my wife and I watched an elk—huge, tall, hornless, almost chalk gray and lean—emerge from the hillocks on the south side of the road. The animal stared at us intently, then crossed the road to join several young elk grazing the deep grasses tucked under the cliff face, where rock meets water: proof of a timeless ecotonal niche that has not yet lost its nurturing power.

Chaco's Wash District

The Chaco Canyon Core lies about thirty-five miles to the northwest of Shabik'eshchee. The core is where most tourists and archaeologists focus their interest. Here, in a tangle of side washes and broken mesas, once flowed the Chaco River—now itself an entrenched seasonal wash. This sector of the Chaco River is lower in elevation than at Shabik'eshchee to the east. That gradient, and several other small, uphill washes, once accelerated water from up-canyon, amplifying the temporary but forceful, and occasionally disastrous, river flows from spotty, scattered rainstorms to the east. About ten to fifteen miles west of the core, any remaining runoff would flow southwest, winding up in Kin Klizhin Wash near the Great House by the same name.

The Chaco Canyon Great House district covered *five to six thousand square miles* as an integrated water catchment system. This water system was enhanced in spots where water velocity could be manipulated by human intervention. Its landform of cliff faces and mesas, stretching about 130 miles east to west and 50-60 miles north to south, trapped flowing water from both seasonal rainstorms and winter snowmelt from mesa tops. Then the region's elevational differences propelled much of that water right past the planted fields maintained by both Chaco Canyon's small house farmers and Great House elites. The water flowing through the wash district was sweeter and less acidic than the San Juan Basin's groundwater, making Chaco Canyon's seasonal waters healthier to drink.

What we do not know is which Great Houses owned the waters of the three great washes, Chaco, Escavada, and Kin Bineola. Nor do we know who tended the crucial up-canyon rivulets or maintained the dozens of tiny, upland water ditches miles away from the Great Houses.

Chaco Canyon communities, both Great House and small house farmsteads alike, did *not* suffer the full consequences of spotty, ephemeral rainfall as experienced by families living in the open, dusty agricultural swales of the southern San Juan Basin, where water catchments were narrow, elongated wash zones of less than a square mile in area. It was only at the larger catchments and wash systems where outlier Great Houses were built in the 800s–1000s AD. To visualize this, rest your hand, palm down, on a table and focus on the bluish veins radiating from your wrist to your fingers—those veins look quite like the map of an extensive wash district.

Chaco Canyon was located astraddle a region where both the Pacific climate patterns (snows, spring rains) and the Gulf of Mexico monsoon patterns (late summer and mid-fall rains) occasionally collided with one another. The washes, mesa edges, hilly side canyons, and mesa foot sand dunes had

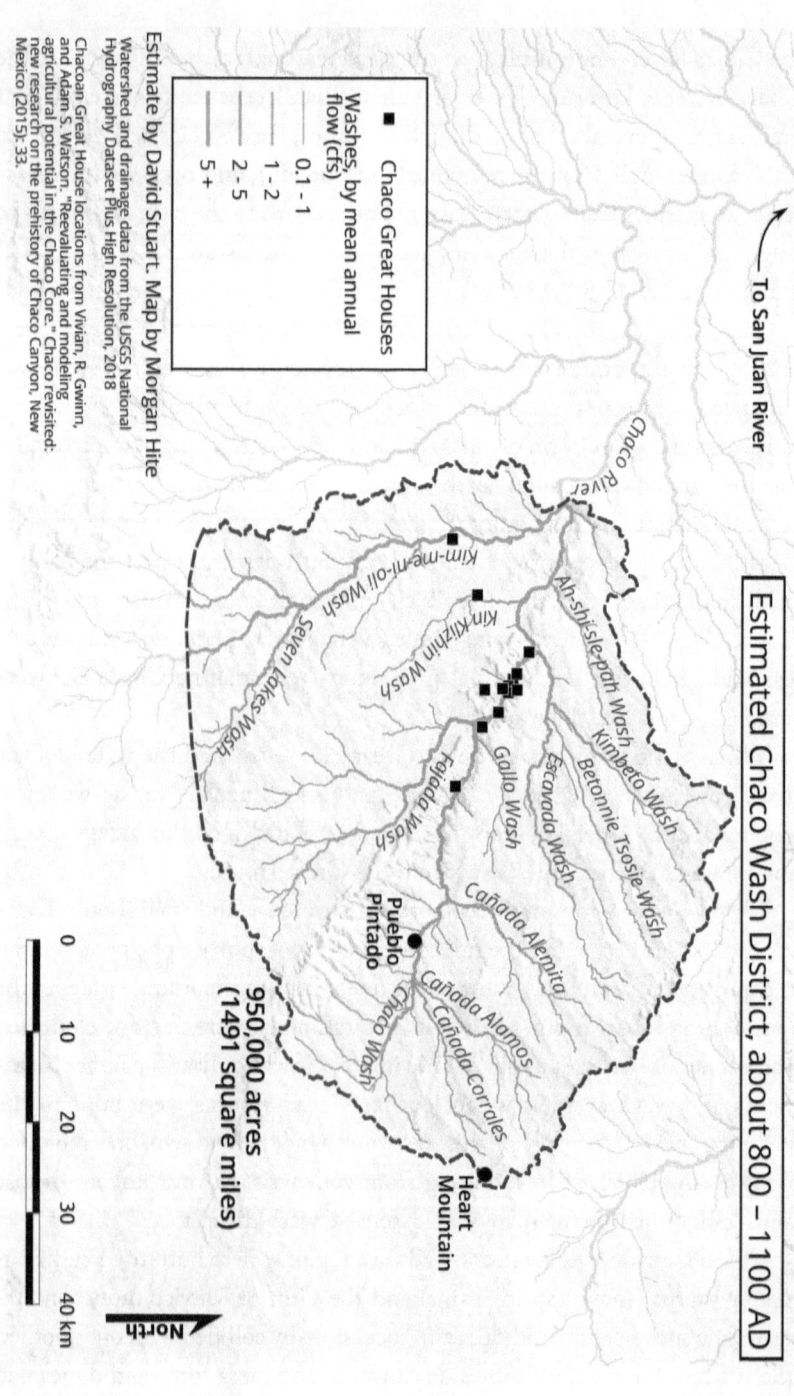

Map 4. Chaco Wash District. Map by Morgan Hite.

all produced corn for centuries. Later, Chacoan-made irrigation channels and stone and wooden head gates were created to slow water, while rock-check dams held the water at the base of the side canyons. This all contributed to a much larger water-flow zone that collected and directed the waters of many spotty rainstorms into coherent paths that flowed into the Chaco River. At one point, perhaps between 850 and 925 AD, Pueblo Bonito had gained enough control of the Chaco River's flow to create a semi-permanent lagoon just southeast of its courtyard, which not only provided drinking water but transformed the very nature of its ancient location to a Garden of Eden that radiated power over nature.

Chaco Canyon's prosperous Great House inhabitants presented as smarter, more successful, richer, better fed, taller, and longer-lived than the struggling corn farmers of the sandy, poorly vegetated swales scattered across the San Juan Basin. The rich in most ancient, complex societies were competitive over their relative wealth rankings. These flashy attributes are the very same that powerful classes in modern societies nurture and project. To this day, the rich of modern societies expend great amounts of money and "cultural capital" to maintain their own financial networks that largely exclude the masses.

Similar power dynamics also likely supported the control projected by the male Great House elites at Chaco Canyon, even though the women were the actual land and house owners. It was the men, however, who modified the modest waterscapes surrounding Chaco Canyon into the above-mentioned interconnected wash district that captured a higher percentage of scattered rain showers and directed much of the water to Downtown Chaco.

Small House vs. Great House Society Becomes Defined

On the outer perimeters of Chaco Canyon's wash districts, several clever water conserving and diverting tactics were utilized. These included cobble and brush windbreaks and modifying ground slopes to direct water flows to corn plots. On Chaco Canyon's south side, most farmers and artisans occupied a mix of both old-style dugout pithouses founded in the 600s–700s AD and the newer, aboveground masonry Unit Pueblos of the 800s–900s AD. Many dwellings were tucked onto the talus slopes of South Mesa. In those north-facing Unit Pueblo settlements, the lack of winter sun required more metabolic calories to retain body temperature as compared to their pithouse neighbors. Thus, both long-occupied pithouses sporting small kivas and an added wing of aboveground rooms was common. Many of these frequently modified households' architecture spanned about three centuries of mixed design styles.

Common folks primarily farmed the sand dunes, which had formed from wind and water-driven erosion of soft rock from South Mesa. The dunes were typically overlaid with denser clays deposited by the uncontrolled Chaco River in earlier times. During summer rains, some water percolated down through moist, sandy layers, coming to rest atop the dense, underlying clays. Varieties of corn that produced long racemes (tap roots) could grow, sprout, and mature in the sand. A nearly identical eco-niche is still used in the famous dune agriculture called *ak-chin* by the Hopi peoples.[9]

As noted above, when heavy rains came roaring down to Downtown Chaco, they erased many farmable fields from the canyon's floor. Thus, constant rebuilding and repairing of houses and field structures became a trait of Chacoan Great House culture. In good years, the Chaco River overflowed its banks, replenishing the adjacent canyon floor with silt and vegetable material. Over time, the canyon's elites appear to have engineered a series of check dams, head gates, and diversion channels that slowed water and reduced the risks of major flooding.

The Great House elites operated at a scale large enough to drive architectural change and enhance institutional and technological complexity. The architecture of their massive dwellings changed far more rapidly than did small house farm architecture. Chaco Canyon's elites were able to launch very complex projects because they could employ and feed large, skilled work crews, frequently expanding the number and size of their ground-floor food storerooms.

Other agricultural endeavors taken on by the elites and their construction crews enhanced water flows and raised crop yields. Such large-scale projects were not available to hundreds of tiny gardening and farming communities scattered along dusty hillsides and small washes throughout the central San Juan Basin. Few small farm settlements could muster the labor needed to reshape the landscape and create a wash system sufficient enough to sustain large cornfields. The best that families could do on the outer fringes of the upper canyon's integrated wash system was to utilize gravity to redirect rainfall to a half acre or so of corn. Many small farmers planted along the margins of the upper Chaco River, along the slopes of lower outlying washes like Kin Bineola, and along the cliff washes in the upper canyons of Cly's, Mockingbird, and Gallo (Vivian and Watson 2015).

Danger: Low Protein and Food Shortages

By the early 900s AD, the effects of insufficient protein and labor pools had begun to limit many rural communities' ability to grow and innovate. Infrastructure did not change much. A few hamlets situated on natural washes or on pockets of richer soil grew in size, but the era of rapid increases in small house communities had stalled. In such hamlets, most work calories were expended just to provide a fragile family life.

By the late 900s AD most small house farmers lived on the edge of disaster. Ordinary small farmers and their children worked to exhaustion, inevitably feeling frustrated, frightened, hungry, and angry. These conditions likely caused children to develop psychological issues such as anxiety, generated by high levels of cortisol and the inevitable changes to their brains' amygdala. Such anxious children would in turn have low problem-solving abilities and fragile social relationships. These factors weighed heavily on farm families' life outcomes in the San Juan Basin.

The Robust and the Frail

Even in the best of planting times, the universal Great House assurance—stored corn—was not a long-term solution. As regional populations grew and groundcover continued to decline, prior solutions to hunger—Indian ricegrass seeds, yucca fruits, prickly pear fruits, wild amaranth, chokecherries, and small animals—failed to meet minimum dietary needs.

Eating more corn did not compensate for hunger, anemia, weight loss, or fragile pregnancies. The medical health profile of a high percentage of the San Juan Basin's growing population declined even when the volume of corn eaten at mealtime increased. Pellagra affected young children and women.

Iron deficiency anemia was widespread, as were dangerously low blood platelet counts. Thinning blood (low white platelet counts) wreaked havoc on immune systems and led to profuse bleeding from even small injuries.

The more small house farmers and their sharing networks struggled to provide food, the more Great House power and influence became solidified in the region. By about 1000 AD, the health and longevity gap had severely widened between elites and the farming classes. This gap loomed large, and grew ominously, in what would be the eventual fate of the Chacoan world.

PART IV

CHACO CANYON'S DOMINION & FALL

CHAPTER 11

MODIFYING LANDSCAPES AS AVENUES TO POWER

BY THE LATE 900S AD, Chaco Canyon's Great Houses had established widespread economic dominion, fueled by roadways to many once-isolated farming quarters. These advantageous trade patterns allowed them to build new outlying Great Houses like Pueblo Pintado to the northeast and the great tower kiva of Kinya'a to the south.[1] In those areas, scattered small house farming populations were still living on severely over-foraged landscapes. Other once-verdant, mixed grasslands had been devastated by long droughts. Hot summer winds, rising temperatures, and blowing dust and sand had erased many of the native plants like amaranth that had once identified good garden locations. Wherever the landscape was still usable for gardening, corn was planted out of proportion simply because dried corn stored well, was useful as fuel, and had steadily risen in market value.

These unstable circumstances offered Chaco Canyon Great House agents the opportunity to establish control over small house planting, ownership, and marketing of large corn crops. I think it likely that the revered Chaco Canyon emissaries combined both heaven-derived planting advice and divination with some aspect of, "If your crop fails, we will make it good by giving you seed corn. No need for you faithful ones to keep more corn than you need to fill your own storage pits. We're here to help the faithful. You do not need to fear starvation. Fill a basket with a share of your crop to our harvest ceremony here in Chaco Canyon. An equal measure of your corn will always be available to you in your season of need. The Gods will be pleased with you."

My above imagined monologue of the Chaco Canyon elites is similar to Franklin Delano Roosevelt's strategy during the sudden and stunning Great Depression of the 1930s. As president, he directed his agents to set up family food banks, soup kitchens, and make-do jobs (build public schools, enhance national parks, create Social Security) and funded culture-reinforcing jobs for school teachers, photographers, folk art makers, musicians, historians, artists, and writers. With the help of his wife Eleanor Roosevelt, he showed up in rural places where no president or senator had ever visited. The Great Depression was no more and no less than an abruptly interrupted power phase based on untrammeled stock market growth that had inflated bogus values attached to publicly traded stock shares. Those bogus values collapsed quickly. Manipulated economic "bear markets" are the presenting, diagnosable symptoms of a corrupt power phase reacting to systemic distress. Such events are numerous in human history.

Social Complexities of the Great House Elites

Massive amounts of investigation and ingenious research protocols have been deployed to determine just *who* the Chaco Canyon Great House elites were. In truth, their precise origins are unknown. Elites may have descended from multiple regions, as the Chacoan world of the 900s–1000s AD is thought to have included several language groups.[2] In addition, a number of Great Houses scattered around the wider Chaco Canyon district did not follow the traditional Chacoan Great House architectural style. Most of these have not yet been investigated in detail.

We do have some insights into the elites' social organization. We know that they practiced primacy of female land and house ownership—it is still practiced today among New Mexico's pueblos—though there was no documentation of the precise female lineages that owned the various Great Houses.

We know, then, that the male elites who lived in the Great Houses did not own their suites (the high priests *may* have been an exception). Their paths to power and influence included ownership of portable wealth, such as Great House storerooms and kivas, and regional and far-flung trade. They had first gained ritual power and status by joining the region's priesthood. A stratum of elite males probably captained specialized infrastructure crews that tended to Great House maintenance, roads, and water movements/control. Men were also likely in charge of the placement of new roadside windbreak rest areas, called *herraduras* ("horseshoes" in Spanish).

It is unknown, however, if the trade of pottery was also dominated by males. Furthermore, the role of pottery making likely had changed hands from women to men at some point. Analysis of fingerprints on bowls and high-quality "gift" pottery shows us that, eventually, males became more heavily involved in the manufacture of higher-end trade wares than females.

While females owned corn plots, their male relatives managed the use and planting of those plots. The business of hiding the true volume of seed, food, and corn in the elites' Great House storage bunkers likely also fell to the men. However, it is unknown who kept account of all these commodities and their values, or how they even "kept the books." Quipu or their equivalent methods remain unidentified.[3] Archaeologists still need to ferret out any tangible hints of a counting system. Was the Great House counting and inventory system a local one, or had it come from afar, such as from the Tucson Basin or western Mexico?

Great House Expansion and Grandeur

Geographic expansion was crucial to the stability of the PII–PIII Chacoan high period (900–1500 AD). But why the push for continual geographic expansion in such a biotically lean regional district? This is no secret—aggregating wealth and controlling the distribution of farm produce was at the core of Great House power. This was utterly predictable from an ecological perspective. Only by controlling the products of ever-greater acreages could they acquire a few more percent of the edible yield from many thousands of more thinly populated farm districts.

In the process of expanding in the late 800s–1000s AD, Great House society also funded and created new roads. Elites expanded and modified outlying wash systems to enhance crop production. This was also the era of grandeur in Chaco Canyon: elaborate, banded-stone masonry work, huge kivas, large religious gatherings, and fabulous and frequently changing styles in pottery.

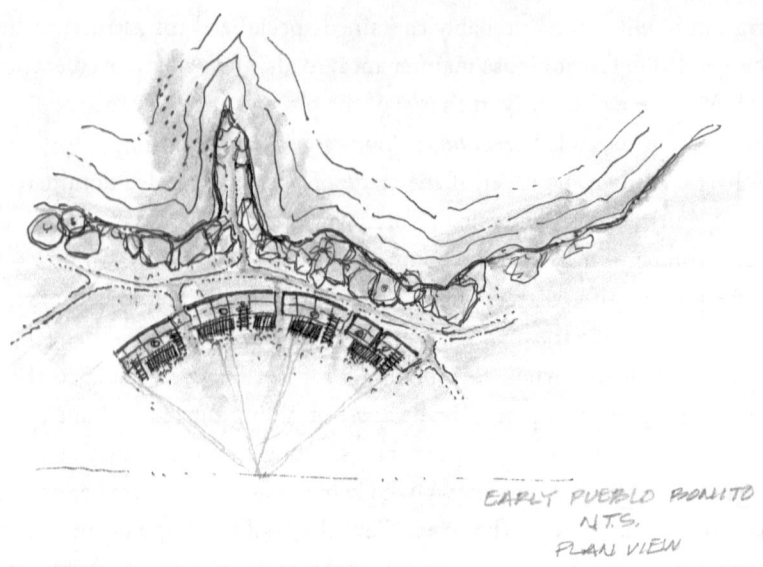

Early Pueblo Bonito plan view. Drawing by Baker Morrow.

Chacoan Great Houses' varied floor plans included squares, rectangular semicircles, upside-down figure Us, squares with towers, and D- and E-shaped Great Houses. I suspect these variations were expressions of the region's complex cultural identities. The most famous of these floor plans is the semicircular Pueblo Bonito, the assumed pinnacle of Great House engineering. Most Great Houses included deep, paired kivas and large, frontal plazas for rituals. Several rectangular Great House floor plans sited near the Chacoan core suggest multiple cultural origins among their owners/builders.

Great House construction episodes peaked at least three times: the 850s–950s AD, a period of below normal precipitation; the early to middle 1000s AD, a period of above average precipitation; and prior to the deep droughts beginning about 1130 AD.[4] Each major episode of public architectural expansions was carefully designed to project images of power, wealth, and control over Downtown Chaco's expanding geophysical and religious domain.

The obvious aim of elites was to utterly dazzle visitors. Great House buildings were earthly proof of Chaco Canyon elites' wealth and power over a vast celestial realm. Several Sun Priest rooms have been identified in upper stories at Pueblo Bonito. Male-focused kiva societies were numerous and ever changing. In short, the spectacle of Downtown Chaco in the ten hundreds AD was unmatched in the region.

Pueblo Bonito's Grandeur

TABLE 6. Building Stages at Pueblo Bonito, Derived from the Work of Bradley Vierra and James Vint (1), as well as Wesley Bernardini (2)

Stage 1. 850s–930s AD: 96 residential rooms, 12 families, many ground-floor storerooms.

Stage 2. 1040s AD: 35 massive two-story room suites added without residential floor features, assumed to be storerooms.

Stage 3A. 1050s AD: A combination of storage and ritual space, as several rooms are connected to kivas. Parrot/macaw skeleton and Mesoamerican objects. Did that stage include cylinder jars and cacao?

Stage 3B. Late 1050s AD: 44 nonresidential rooms added.

Stage 4. 1060s–1075 AD: Block of 18 ground-floor rooms added. The first two rows were one story; the third row was two stories; the fourth row was a full three stories. Some of the room suites housed elites. This part of Pueblo Bonito faces southwest towards the Chuska Mountains.

Sources:

(Vint 2005)

(Bernardini 1999)

By the late 800s AD, Pueblo Bonito had become the primary focus of Chaco Canyon's commercial and religious activities. It faced south by southeast, stood four stories tall at its rear, included truly ancient rooms at its right rear quarter near the cliff face, and housed several lineages of Chacoan females. The elites were not particularly numerous—estimates of Great House population numbers vary among scholars, ranging from twenty-five to seventy-five resident elites, male and female (Bernardini 1999).

From a pathway in South Gap, visitors arriving in late summer would have seen Pueblo Bonito's stone walls and terraced habitation rooms near the cliff face that rose just behind (in another century, the facade would be four stories). In the foreground, they would also have seen numerous corn plots, some squash and beans included, tucked up along the sandy areas on the north-facing slopes at the base of South Mesa. These plantings ranged in size from 600-800 square feet to a quarter acre. Many tiny fields were defined by their windbreaking cobble and brush berms, which fended off spring's cold west-to-east winds.

A traveler entering the canyon through South Gap would also have seen views of both Unit Pueblos, a number of ancient pithouses, some attached to small, aboveground masonry Unit Pueblos. Crossing the canyon toward Pueblo Bonito, the traveler would also have passed narrow irrigation canals, larger cobble and log irrigation head gates, and many surrounding cliff base dune fields dotted with small corn plots. These numerous small house sites in Chaco Canyon would have been familiar to visitors, as most of them looked like their own rustic hamlets. The only large acreage farms in the region were near present-day Newcomb, New Mexico (Friedman, Stein, and Blackhorse 2003).

In contrast, the wonderment of visitors would have been energized by the three largest early Great Houses: Pueblo Bonito, Peñasco Blanco, and Una Vida. We know much about Pueblo Bonito, but very little about the other two. Shortly after the University of New Mexico Regents transferred ownership of Chaco Canyon to the National Park Service in 1949, the federal agency closed most of Chaco to excavations. What DNA types predominated among residents of Peñasco Blanco and Una Vida? What pottery types lie fractured under the floors?

Recall that Pueblo Bonito's first large construction phase dates to the 850s AD. In the Great Houses, residential rooms were distinct from both storerooms and religious spaces. Bernardini's research discusses Bonito's growth phases and the patterns of door connections and floor features that distinguished residential rooms from storage space (Bernardini 1999). Residential rooms were stacked above ground-floor grain bunkers. Access to upper-story family rooms was typically gained by way of ceiling openings and wooden ladders—quite secure.

In contrast, storerooms tended to be side by side and on ground floors, just as in medieval Europe. Ground-floor storerooms reduced lifting/carrying labor, were cooler, and offered no public view of family residential space. Other interior spaces included smallish kiva-like structures. Public life took place on roofs and terraces (likely female focused), in the open forecourts fronting room blocks, and in large kivas (likely male focused)—distinct forms of landscape space. But daily life that was open to visitors' view took place in Pueblo Bonito's grand forecourts. This is where pilgrims were received. Were there special spaces for visitors? Where did the visitors sleep or eat? Was it mostly men who visited? Or did entire families come as pilgrims? Many details of Great House life remain unanswered. Time, dust, winds, a falling cliff face, rotting beams, and loosening stones have all taken their toll at Pueblo Bonito. Were it not for the district's Navajo stone masons, especially the very skilled Huerito family, many parts of Pueblo Bonito that are currently open to visitors might not still be standing.

In any case, Pueblo Bonito's first large expansion in the 840s–850s AD was both soundly engineered and beautiful. Before the first expansion phase in the mid-800s, no small farmer could have imagined the eventual scale nor the massive labor pool required to build Pueblo Bonito. Over time, many of their sons, grandsons, and great-grandsons might have worked on the labor gangs that built roads and added more upper stories to sites like Pueblo Bonito. It was an excellent approach, but not perfect. Each new story added to Pueblo Bonito was more expensive, calorically and material-wise, than the story below it. It cost millions of additional food calories just to lift hundreds of tons of shaped stone to a greater height. The greater heights also suggest the need for wooden scaffolding, a valuable commodity in its own right, as were the yucca fiber baskets probably used to lift stone blocks and tons of clay needed to finish the walls.

Cornfields Reshape Daily Life Among Ordinary Families

As noted in sidebar 19, borrowed from the second edition of *Anasazi America* (Stuart 2014), an acre of corn planted in rows, often thinned out during the midsummer growth cycle, could support 540 cornstalks. This produced about seventy-one pounds, or 13.4 bushels, of husked cob per acre. Such acreage would have been tended by male growers who worked the large plots. Some men probably added stone grids to the fields to capture water flow and built cobble/brush windbreaks to protect the young plants from spring's cold west to east winds.

Larger fields supported larger families. My estimates, based on my own

EXCERPT FROM ANASAZI AMERICA (STUART 2014), PP. 102-3, TEXT BY JENNY LUND SHERMAN AND EDITED BY DAVID E. STUART, 2024

Traditional Hopi field plantings typically include the Three Sisters of maize, beans, and squash. These are planted in clumps approximately nine feet apart. Thus, a modern acre measuring 212 square feet on each side could theoretically contain 540 plant clusters, but corn quickly depletes soil nitrogen and native farmers ordinarily move rows each year—planting between the previous year's rows in fresher soil. That practice removes one potential row at field edges. So we assume about five hundred clusters as average in actual acre plantings.

As the prehistoric Four Corners population grew and crowded the landscape, corn rows were planted at closer intervals to temporarily bump up yields. This very likely also happened during extended droughts, when tightly spaced crops were planted in wet fields (stream or spring fed) because more food from the most dependable fields would be needed to sustain Four Corners families. Because of nitrogen depletion, such emergency strategies only worked in the short term.

An average Four Corners farm family of eleven members at about 400 to 800 AD would have needed to consume about 21,000-24,000 calories daily. Increased levels of duration or intensity of work activities would have required more daily dietary calories. Fluctuations in first and last freeze dates, annual precipitation, and storm events all affected the harvested calories. We used published scientific data to estimate the average caloric content that an acre of corn produced under different ancient precipitation scenarios.

Research by correlated dendrochronological records (tree-ring growth) estimated annual crop yields over twenty years in the prehistoric Southwest and calculated bushels per acre of maize at a fixed (dehusked) weight of 31.5 kilograms per bushel and a mean of 13.4 bushels per acre (between 650 and 800) (Burns 1983). From this, we created an equation to determine the average caloric output (corn only) per one acre of arable land.

We took the mean number of bushels (13.4) from our twenty-year span, multiplied it by 31.5 kilograms (the fresh, dehusked weight of one bushel of maize), multiplied by the dry kernel weight (74.4 percent) per cob from Adam et al.'s study on historically different maize varieties (2006), and finally multiplied by the number of calories in 1 kilogram of corn (3,600) (Convert To). This equation showed that if all else went right (no major hail storms, flash flooding, grasshoppers, or the like) an acre of decent, dry-farmed land between 650 and 800 AD could yield 1,168,812 consumable calories of maize.

From this calculation of caloric yield per dry-farmed acre of maize and the daily caloric requirement of a family of eleven based on basal metabolism, body weight, and age (from archaeologically analyzed skeletons and ethnographic data), we can determine how much maize a family would need at any given percentage of their total diet. Published sources estimate that prehistoric Southwestern diets include 65 percent to 75 percent corn. At 70 percent dietary maize calories, a family of eleven would need to plant about five acres to produce the equivalent of 239 modern cobs of corn daily (we adjusted for smaller prehistoric cobs as measured in archaeological reports). That translates to about twenty-three cobs of corn, in all its forms, per person per day.

By 1050 AD, larger cobs and better water control increased an acre's yield, especially if the rows were shifted and the fallow rows were planted in beans or amaranth to replenish niacin in the soil.

fieldwork among Indigenous farmers in Mexico and Ecuador, suggest an extended family size of ten to twelve members, including in-laws and the aging, who lived in side-by-side in stone, mud, and wattled Unit Pueblos averaging about eight hundred square feet each. About three hundred square feet of that was reserved for storage. Additional living space included rooftops, *ramadas* (sun shelters), and open plazas fronting a row of Unit Pueblos. In winter it was smoky and dark inside; in late spring they lived in their patios and courtyards. Compacted dirt plazas included many corn-grinding areas, exterior fire pits, and several food storage pits.

Many of the larger multifamily farm hamlets also had a semisubterranean community house. Archaeologists argue incessantly about the precise usage of these deep underground rooms, suggesting winter quarters for unmarried males, as well as religious events. It is safe to say that these subterranean chambers had multiple uses.

The Powerful Sense of "Belonging"

As noted above, Chaco scholar Nancy Akins's research on human bones found at both Chaco Canyon's small farmsteads and the larger, multiroom Great Houses makes it clear that the small house people in the Chaco Canyon district surrounding Pueblo Bonito worked both longer and harder than residents in the nearby Great Houses (Akins 1986).

By the late 900s AD, the small house farmers living within sight of Great Houses like Pueblo Bonito, Peñasco Blanco, and Una Vida experienced very different lives than did the Great House elites—the San Juan Basin world was not an equal place for this struggling underclass of farmers. Their struggles, in fact, cannot be overstated. Population was rising and their economy was fragile. Their storeroom capacity had stagnated while Great House storerooms expanded. Small house families endured frequent bouts of hunger and infant deaths. Mothers often could not properly nurse their infants due to the ravages of iron deficiency anemia, pellagra, hard labor, and the lack of daily protein. An infant death wasted an investment of the quarter million calories consumed by a woman during her pregnancy and the first six months of nursing (Stuart 2014)!

Yet if one star-crossed male could not live the envied lives of the elites at the top, he could at least bask in the glow of the place for a few days annually—days in which the human ego could be massaged into the illusion of "doing better" and "belonging." Small house farming men were likely elated to be formally received by high status male priests, who were the representatives of both a heavenly and terrestrial empire. The priests laid claim to a much higher status than even the ancient female lineages. Their status gave them agency. Thus, for a rural, small house male to be included in a Great House kiva's religious ritual not only meant acknowledgment of belonging but a potential path to economic success with the opportunity to trade their bulk corn.

I envision the priestly elites inviting male guests to convey some version of a "corn tithe" offering to a kiva society in exchange for a promise to receive excess corn from the Great House storerooms. Elites may also have promised dried corn for the faithful to carry home in seasons of drought or famine—all the newcomer had to do was expand his cornfield and deliver a portion of his crops to the Great Houses. Great House promises of dried or seed corn in a bad year would have quickly reduced the number of an ordinary corn farming family's traditional sharing-network obligations. That benefit meant that corn farmers could keep a higher percentage of their own corn crops.

Risk Versus Reward in the Corn Economy

Sharing their corn with Great Houses like Pueblo Bonito generated new risks for small house farmers. Yes, storerooms at Chaco Canyon *could* provide emergency supplies of life-saving food and seed corn and, thus, provide a needed sense of security. Corn had become a core asset and investment with a rising value, just as tobacco was a monetized asset in America's Chesapeake Bay region between the 1680s and 1800s.

Yet, as revered as corn was, it was *not* everything. It was a crop that could be severely limited by precipitation, and it rapidly diminished plant resources, groundcover, and vital nutrients in the soil. Corn was stingy nutritionally, and the Great House stores did not provide a solution to the dangerous, long-term effects of a corn diet.

By the 900s AD small house farmers needed to enhance their long-term food security and prepare much more diverse diets. As a result, small house populations struggled to maintain local family sharing networks that had long been the mechanism that moved healthy foods from one ecological zone to another, such as piñon nuts, tepary beans, amaranth, sunflower seeds, chenopod leaves, and turkey meat.

Meanwhile, built on others' surplus garden products and expanding cornfields, Great House wealth grew. Elites' lifestyle grew more elaborate and costly. Staging and celebrating large groups at solstices, moon phases, and elaborate prognostications of the region's seasonal trends all cost precious resources. Rising shortages of those resources appears to have triggered obsessive Great House desires to connect to and own an ever-growing interaction sphere. That need may have shaped the late 900s AD as road systems expanded.

In reality, the Great House Chacoans of that time had invested too large a share of the region's essential resources in expanding their own costly infrastructure and subculture. They had become comfortable with their power. In some cases, those infrastructure projects were beneficial. In other cases, a project cost more to fuel human labor than gained from bulk corn brought to them from a dry, distant corner of the San Juan Basin.

The scale and nature of trade fluctuated with the patterns of the rains. Over time, many resource strategies were explored. Some worked, like the spread of domesticated turkey flocks in the 900s–1100s AD, the meat and eggs of which enhanced the needed animal protein in ordinary folks' diet. Native cotton was also farmed, as its seeds produced calories and dietary oil. In short, daily life required a fragile and complex game of ecological and demographic chess—a game in which the values of the ecological chess pieces rose or fell with the region's capricious rains.

CHAPTER 12

THE RHYTHMS OF GREAT HOUSE POWER, 900S-1100S AD

A PERIOD OF BELOW normal precipitation during the 900s and early 1000s AD saw Great House construction proceed in bursts, followed by substantial periods of diminished construction activity. We do not yet know if most major construction episodes were tied to climatic events or to a celestial or religious cycle. In any case, Chacoan bursts of Great House construction were intermittent like a pulsing star.

The Red Mesa Valley District

According to John Kantner and others in Frances J. Mathien's monograph, *The Casamero Community in the Red Mesa Valley of Northwestern New Mexico*, the Red Mesa Valley district exuded an air of independence from the rest of Chaco Canyon (Mathien 2010). Its Great House Casamero and extensive wash basin morphed into large village complexes that aped Chaco Canyon style on the cheap.

Apart from looks, however, the district remained somewhat separate from the activities and trends going on around it. Sometime in the late 900s or early 1000s AD, Casamero's wash system was upgraded to further enhance water flow and direct it to large corn plots.

Casamero occupied an area replete with impressive red sandstone mesas, footed by two modest rivers and numerous seeps and springs. The Red Mesa Valley encompassed the most valuable real estate in the southern Chaco world, with access to both Mount Taylor and the Zuni Mountains where big game, turkey, deer, pine, and roof beam construction timbers could be procured.

Construction Episodes

As noted in the second edition of my textbook *Anasazi America* (Stuart 2014), major construction episodes at central San Juan Basin Great Houses came about every twenty-five to fifty years. Late period Great Houses (1050–1130s AD) were built in far-flung districts like the southern San Juan Basin, west-central New Mexico, and eastern Arizona. Many of these late period outlying constructions were architecturally less detailed, not built to the same quality as the core Great Houses strung along the Chaco River and its washes.

Major construction events may have coincided with periods of more stable and predictable precipitation or religious calendrical cycles. They may also have occurred as roads in Chaco Canyon were extended, as suggested by archaeologist Thomas Windes. The most reliable Great House dating has been due to Windes's work collecting and tree-ring dating Chacoan era wood. His data suggest several distinct eras when Great House and other large stone structures bearing Great House features were built at the same time as road expansion (2003).

Altogether, about fifty-six Great Houses were built over the span of 350 years. Not every Great House was identical, nor was each one's purpose the same. Distant Great Houses may have housed Chaco Canyon clans and a local lineage of several southern Mexican trading entrepôts. Others may have been outlying storerooms built to enhance the speed with which the elites' emissaries could distribute dried corn in a remote district. Scholars like Barbara Mills believe that the Great Houses probably perpetuated important family lineages. Indeed, the multistory residence suites suggest that the rank of individuals living inside of them must also have been tiered. Great Houses also solidified architectural memories of seminal events and expressed religious power over nature and land/resource ownership (2015).

Clearly there were differences in scale and wealth among Great Houses. Mills highlights several striking construction details found at Pueblo Bonito that were not found elsewhere, such as finely joined log pilasters and wainscoting, which likely provided insulation due to the use of bunch grasses. Mills also notes that of those fifty-six known Great Houses in the San Juan Basin proper, seventeen did not have a great kiva (2015).

The Enigma of Pueblo Pintado

As the Chacoan Great House world increased its interactions with more distant basin communities through waves of new, outlying Great House and road construction between 900 and the 1100s AD, the Great House of Pueblo Pintado stands out as a mystery. Built in the 1040s–1050s AD about thirteen miles east of Chaco Canyon's washes, its purpose remains unknown. According to Michael Marshall, John Stein, Richard Loose, and Judith Novotny—some of my generation's finest and most astute field archaeologists—Pueblo Pintado was characterized by great construction skill but unexpectedly sparse evidence of habitation (Marshall et al. 1979). Yes, some of its small valley-end dune fields appear to have been farmed, based on artifact finds, but there are virtually no daily artifact finds (cooking bowls, etc.) in or near the huge room blocks to indicate that anyone actually lived there.

Pueblo Pintado was brilliant architecturally speaking—its paired kivas clearly echo the architectural layout of Pueblo Bonito. But how it delivered on the return of its huge investment is not known. One of its possible uses could have been to support the families who tended the upper canyon's many small acequias. Dropseed and other bunch grasses used for storeroom floor insulation and sacred burials were harvested in this area. Indeed, a number of documented models of such Indigenous "water keeper" communities do exist elsewhere in the arid regions in the Americas (Bennison 2023). How much support and protection the Great House gave this upper canyon wash area merits more research.

Pueblo Pintado seems to have served as a very costly icon. Indeed, by 1040 AD, the Chacoan Great House world was rich enough to make bets on enterprises like Pueblo Pintado that might not pay off in corn or trading power. Was Pueblo Pintado a "make work" project designed to maintain the loyalty of a cooperating farm district's hungry labor force? Was it a project designed to award paying work to the specialized and highly skilled

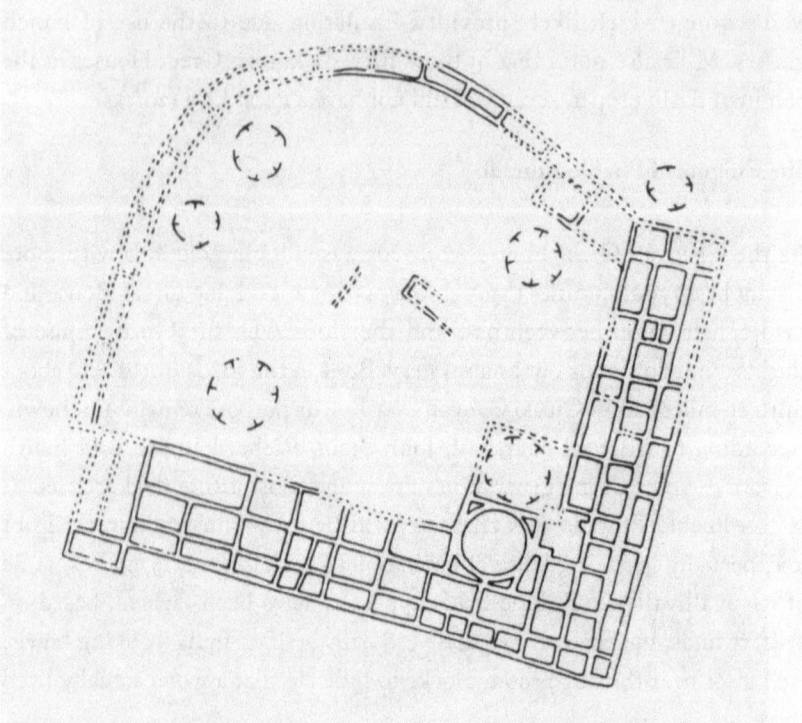

Figure 7. Pueblo Pintado plan view.

water-flow construction teams that Chaco Canyon supported over several centuries? Or was it actually designed as a territorial barrier? Did it act as proof of Chaco Canyon territorial propriety because of its extensive plantable dune fields, and/or assault protection from the aggressive and reclusive Gallina Mountain populations to the east? Pueblo Pintado may have been built in response to a blend of all these factors.

Ownership Patterns Between Men and Women

Recall that men appear to have owned moveable and transportable property, with the exception of their kivas, and women most likely owned nonportable property, like real estate. San Juan Basin women may have also owned storerooms, corn plots, and forage territories, as scholar Elizabeth Chestnut has discussed with me during many conversations in 2023.

It seems logical that men played a large role in the transformation of a large, one-to-two-acre outlying corn plots by adding stone berms, leveling and redirecting the landscape to control water flow, and removing unwanted rocks. Their reward was ownership of an unknown portion of the corn they harvested—perhaps half or a quarter's worth.

Many important questions about the ownership and distribution of crops remain unanswered. Was the men's portion of harvested corn marketed locally, while husked and tasseled corn was deposited in Great House storerooms? Was there both *sacred* corn (tassels and unhusked, pollen intact) and *secular* corn (husked, without tassels and pollen)? Just who owned the bulk corn before it reached a Great House storeroom, and who owned it once stored? Who owned the portion of a corn crop that reached a small house woman's rear storeroom, and how did that contrast with ownership and control of corn stored in a community's highly visible hillside corn bunkers? Did patterns of ownership change over time? These are important questions to be answered.

Power

Edible corn was equivalent to money and, thus, potential power. Humble sharecropping could have led to modest power among those small house males who excelled at producing large corn crops. Modern socioeconomic analysis tends to view sharecropping through the lens of freed slaves and poor, landless whites during the American South's early post-slavery era. But sharecropping corn in the San Juan Basin consisted of a woman granting her male kin permission to farm a portion of her inherited land in return for a share of the crop.

Meanwhile, the Great House source of socioeconomic power seems to have come from the aggregation of huge quantities of stored food, sun and moon cycles, and the religious prognostications of the Great House priests. A second basis of power came from its ancient primacy of land ownership, still revered almost everywhere.[1] Its third focus of power came from the engineering of its interconnected water flows in the core wash region of

Great Houses and their storerooms. Lastly, its fourth source of power derived from roads, rest stops, the engineering of numerous distant wash zones, like Casamero in the southern San Juan Basin and, finally, Pueblo Bonito's ancient connections to Basketmaker/Mesa Verde culture and Mesoamerican trade networks.

As population rose in the 800s–1000s AD, the value of stored Great House corn rose dramatically. It is likely that only a few people knew just how much corn was actually stored in the Great Houses at a given time. Just how old was the displayed corn in a Great House bunker during a religious gathering at Chaco Canyon? Seed corn typically has a storage life limit of six to eight years, but dried food corn can be ground and eaten even decades later.

Pueblo Bonito's Apogee

Pueblo Bonito had, at great human labor and environmental costs, reached its architectural apogee in the 1060s–1070s AD, during its final building stage (see table 6).

Recall that Bonito's most ancient rooms were below ground; the newest rooms were the highest yet, of four ample stories in height. For its last construction phase, a block of eighteen single-story large ground-floor storage rooms were added. Stacked atop these were more two and three-story, multiroom suites to accommodate more elites who may have abandoned outlying Great Houses. As a finishing touch, a suite of rooms were built, facing southwest toward the Chuska Mountains. This fits with the widely published archaeological assumption of Chaco Canyon's long-term alliance with the Mesa Verde and Chuska Mountain communities, Chaco Canyon's prime source of structural timbers.

Archaeologist Stephen H. Lekson has published the computed labor costs to build out Pueblo Bonito's final structure of four stories with a roof terrace and hundreds of new rooms (2006). What Lekson never fully explored in his doctoral dissertation on Pueblo Bonito construction was that the stones carried up four high stories to complete its final roofline in the 1060s AD cost hundreds of thousands—if not millions—more work calories to lift and place upper-story stone than the early AD 850s additions of ground-level storage and second-story residential rooms (Stuart 2012). Pueblo Bonito's fourth story and huge rooftop terrace required temporary scaffolding, huge quantities of yucca fiber rope, and large crews to set and chink thick upper walls. As noted above, this could be viewed as a budget-breaking marker of problems to come for Chaco Canyon.

While the construction data from Pueblo Bonito are messy due to lengthy building stages, reused structural timbers, and degraded remains of everyday life, there are other details we can glean. This final building phase would have been initiated after enough materials had been gathered and prepared for use. It might have taken years, if not decades, to collect, cut, and "purchase" structural timbers in the nearly treeless central and southern San Juan Basin. Connections to other communities living in forested districts far from Chaco Canyon were essential to Pueblo Bonito's needs.

It cost millions of dietary calories to carry and shape construction stone for banded Great House walls like Bonito's. A male weighing about 125 pounds could only carry about 50–60 pounds over a long distance. Thus, this single planned building event at Pueblo Bonito could easily have spanned several human generations, including the years gathering materials. One man with a life span of about twenty-five years could have carried and rough-shaped loads of stone that his son, or even grandson, fit into place by fine-chipping them, then inserting into Pueblo Bonito's final upper story outer walls.

Great House projects like that of Pueblo Bonito's final building event were almost certainly scheduled to match the seasonal availability of male labor. An ordinary male's farming year was intense in spring (planting), moderate in summer (weeding, trimming, and thinning corn rows), and intense again during the fall (harvesting and husking). Men were likely recruited for Great House labor in between their farm work: late spring, four or five weeks in midsummer, and four to six weeks in late fall.

After the hired men had carried, shaped, and stacked stone, acquired and shaped the necessary beams and structural timber (from great distances), rotated the timber regularly in order to maintain straightness, they would amass a supply of dry grasses and lignite. Dry grass was integral to the construction and finishing of the Pueblo Bonito's inner walls, and lignite was essential to its foundations in clayey soils. Newly tamped and smoothed clay storeroom floors were most likely overlaid by a thick layer of dry ricegrass or sand dropseed grasses. Most of the ground-floor storerooms were thick-walled, protected from sunlight, and much cooler than upper story rooms.

The banded stonework of Pueblo Bonito's latest era was of extraordinary quality—an architectural signature of its overall grandeur. Bonito's brown banded walls must have looked stunning as viewed under the canyon's turquoise skies.

The Status Quo in the Ten and Eleven Hundreds AD

In Paul Minnis's published table "The Most Critical Characteristics of Chaco" (2015), his top three Chacoan cultural characteristics all point to power-focused traits. These included, first, monumentality and a managerial class. Second was the included concentration of ritual-based economic power and authority in the hands of competing corporate groups, which focused on corn, ritual, and labor. I translate this as competing female-based clans/lineages in economic competition with the male priesthood. Third, they included the ability of leaders to mobilize a large labor force on a massive scale for Great House construction, operations at the wash district, and road projects. These characteristics all generate, concentrate, and organize economic, political, and religious power (Minnis 2015).

With that power, wealth, status, and the life-saving storage of food, Great House elites projected the vision of a beneficial, food-secure realm that, in reality, actually worked on much smaller margins of stored grain than district farmers knew. For instance, each movement of corn invited a local "rake." One hundred pounds of dried corn delivered to Pueblo Bonito may have become eighty pounds of corn delivered to Kin Bis sa'ani, which became sixty pounds of dried corn given out to farm families.

Furthermore, male farmers delivering corn to a Great House may never have grasped the possibility that the elite men they met did not actually *own* their Great House, even if they might have owned or controlled the storeroom's contents. From the elites' perspective, it was crucial that small house farmers see their huge baskets of corn disappear "safely" into Great House

Photo 11. Bis sa'ani (house atop clay) sits atop a dramatic clay and shale pinnacle about seven miles northeast of Chaco Canyon. Built in the 1130s, it contained the latest roof beam found in any Bonito-period site (1139) and may have been built to protect the central canyon's great houses from unrest among Chaco's northern farming communities at that time. Virtually all of the dark rubble cascading down (*right center*) consists of tabular sandstone blocks once part of the citadel's walls. (Courtesy NPS, Collection 0002/043.004, Negative 12707.)

storerooms, and equally crucial that their priests keep their status and agency of predicting climate cycles.

By about 1050 AD, the elites had pushed beyond their regional landscape's ability to restore itself. The amount of labor and natural materials used to expand and support Great House society and its trading aspirations created an increasingly hungry, angry, and shorter-lived rural farming population. Some Chaco researchers suggest that a 30 to 40 percent infant mortality rate might have been normal by the mid-1000s AD.

Meanwhile, Chaco Canyon's wash district continued to expand. By the late 1000s AD, somewhere between fifty and sixty Great House structures dotted the San Juan Basin. This included the Mesa Verde district, the San Juan and Animas River basins, the Chuska Mountains, the Lukachukai highlands, and an increasingly powerful southern rim of Pueblos like Zuni, Acoma, and Hopi. Most of these ancient settlements had access to groundwater and timber. Thus, they rose in influence as the quality of life declined in the desiccating central San Juan Basin.

In their highly productive gardens, ancient women had been the first to create a diverse and balanced diet rich in niacin and iron. As population in the San Juan Basin grew in the ten hundreds AD, more corn was required and planted at the expense of protein-rich beans and squash. The deep adaptational flaw of this choice was already starting to show: a decline in nutrition had led to ever more cases of pellagra. Pellagra would go on to play a significant role in the disintegration of Great House society during the 1100s AD.

Yet despite the nutritional shortcomings, regional population continued to increase. Demography, the study of population dynamics, is both complicated and intricate; here, I rely on the works of Nobel Prize winner Robert

Photo 12. Traditional dancers at Acoma atop its high "enchanted mesa" about seventy miles west of Albuquerque, about 1898. Note the multistory house blocks, ladders to roofs, and contrasts in dress—traditional women's dance garb but manufactured clothing worn by male spectators along the rooftops. The cut branches are part of a shrine, similar to the brush bower near the kiva in the earlier scene at Oraibi. (Courtesy NPS, Collection 0028/006.002, Negative 77619.)

Photo 13. An Acoma drummer in a flat, crowned hat cuts a majestic figure at a feast-day dance, about 1898. Note the combination of traditional and European clothing on Acoma men in foreground and left—Levis, European hats, and store-bought shirts—and contrast with the Navajo blanket (bottom center) and buckskins (right of blanket). Here, too, the Acoma women (right of drummer) are more traditionally dressed. In right background, three women in Victorian dress (one with parasol) are tourists. (Courtesy NPS, Collection 0028/006.002, Negative 77625.)

W. Fogel for guidance. If I understand the charts and tables in one of his later works, *Explaining Long-Term Trends in Health and Longevity* (Fogel 2012), it strikes me as nearly impossible for the original San Juan Basin population to have increased eight or ninefold between 600 AD and 1100 AD! Substantial in-migration by other people must have been part of the Great House world's demographic equation.

As the population continued to grow, so did the socioeconomic gap between Great House elites and small house farmers. Great House food-producing efficiencies of scale were never fully shared with the general farming population. While the Chaco Canyon road system was an efficiency shared with the basin's farmers and elites alike, those roads likely carried far more resources and trade goods *into* the Great Houses than were distributed *outward* to small house farmers in hard times.

Photo 14. Taos Pueblo about 1898. This view focuses on the adobe beehive ovens used to bake bread—a Spanish influence after the introduction of wheat. Most contemporary photographs of Taos show trees and vegetation. Nearly all these Pueblo scenes in the 1890s are barren for the simple reason that food was cooked and homes heated with firewood. Nowadays, most Pueblo homes, including those in Taos, are supplied with propane and electricity. (Courtesy NPS, Collection 0028/006.002, Negative 77825.)

We know this gap in well-being existed based on the significant differences in height, longevity, and general health between accidentally exposed human remains of Great House elites and small farmers. In short, the projection of a magico-religious Great House way of life primarily enhanced the lives of a few hundred elite Great House dwellers at the expense of thousands of small house populations.

CHAPTER 13

TIPTOEING ON THE EDGE OF CHAOS, 1130 AD

THIS ERA OF CHACO Canyon's history was brutal and tragic. By the twelfth century AD, Great House culture and its corn-filled storerooms had long since replaced the earlier multigenerational family sharing networks. By inserting themselves as the nexus of trade, security, and salvation in times of drought, the Great House elites had supplanted and modified the very fabric of family lineages, local, village-based support, and "empathy" obligations to one another. Expressions of empathy are highly shaped by cultural, religious, and status realities. In times of catastrophe, empathy is typically focused on close family and friends. Great House/small house obligations had worked in tough times and in short-term bursts, but even the Great House elites proved useless in late summer of the year 1130 AD, when the monsoon rains did not come anywhere in the region.

The Hammer of Fate Falls on Chaco Canyon

A multi-decade drought beginning in about 1130 AD unwound the fabric of San Juan Basin society. Many dryland farmers lost their corn crops, and the region's small house population turned to Great House food storage for help. The Great Houses had, for centuries, targeted their food storage to last for three to five year cycles of corn growing and storing, so long as seasonal rains kept up. In the absence of rain, only the massive Great Houses could provide enough corn to stabilize the diets of thousands of families. But Pueblo

Photo 15. Salmon Ruin great house on the banks of the San Juan River (*out of view, right*) about seven miles southwest of Aztec. Note the same C-shaped room block (*left*) with kiva in courtyard. Founded just before 1100, Salmon was a Chacoan refuge until a number of its women and children were burned in the tower kiva that once arose from the main block (left center). Modern buildings in the far lower left are from a field school held at Salmon in the 1970s. Replete with massive quantities of burned / dried corn, Salmon Ruin's huge food reserve suggests that some Great House elites expected to return from their hegira in southern Arizona to the San Juan River country. (Courtesy NPS, Collection 0002/043.004, Negative 18007.)

Bonito closed its outer gates, erected several protective stone walls, and deployed bowmen to protect nearby road segments and create lookouts. Within its walls, elite residents clung to their bunkered corn.

Once cut off from Great House support, small farming hamlets that had been entrenched in the Great Houses' closed system of trade were truly in dire straits. Three centuries of social compacts that had stabilized fragile regional populations were ripped away as ordinary folk realized they were being abandoned and ignored.

Starvation followed, and fueled rage. Evidence of widespread violence, burials, and burned hamlets has been found in many districts. In the northwest, Mesa Verdeans abandoned many of their large, mesa-top villages and fields and moved into more secure cliff palaces.

Great Houses were forced to look farther outward than ever before, seeking to trade and command influence over an immense region that included southwestern Colorado, the region of Utah's ancient Basketmaker cliff dwellings, and even northern Mexico. As the wealth of the Chacoan core flatlined, local farms failed and profit margins diminished. Hoarded corn became an obsession. There were no new nearby wash districts to create. There were no more giant conifer trees to cut for roof beams. The latest road built in the 1100s AD, one labeled the "Great North Road to the Underworld," was likely a "make work project" to engage restless men and reaffirm ties with Great Houses in the northeastern San Juan Basin. Their geographic expansion resulted from desperation rather than comfortable dominion.

Three centuries of irreversible ecological damage had been done in the Chaco Canyon heartland. Tens of thousands of square miles of highly diverse piñón-juniper ecotones had already been stripped bare, returning to sand, coarse clay, and scant desert shrub. With that came the loss of rabbit, migratory birds like geese, piñón nuts, juniper berries, and lizards. Even desert plants like yucca, once so valuable for cordage, sandals, and baskets, had diminished due to overharvesting.

Some of the northern Great Houses and large farm hamlets responded to this by growing more native cotton. Cotton seeds provided edible oils as well as fiber—both essential to daily

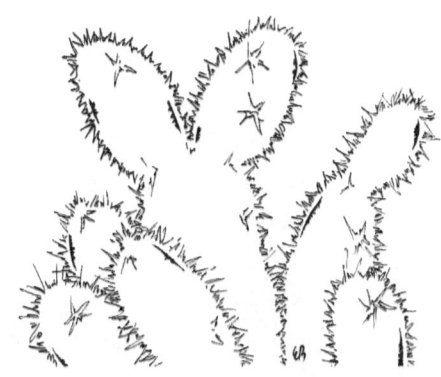

Barrel cactus and cholla sketches by Esther Burton.

Native cotton sketch by Esther Burton.

life. They also refined the techniques of turkey husbandry, and increased ak' chin corn farming. These responses were successful enough that some stable Puebloan communities far from the central canyon did not rapidly lose their core institutions. Meanwhile at Chaco Canyon itself, the facade of sacred ritual could not hide the onset of social and economic disintegration.

The Edge of Chaos

The massive losses of social, economic, and ecological Great House stability combined to approximate the human systems labeled as being on "the Edge of Chaos" by M. Mitchell Waldrop in his book *Complexity* (1992). In Waldrop's analysis, a complex human cultural system that had once attained a balance point but in which the major components never quite locked into place often degrades into a fragile, negentropic state. Such a turbulent state constantly shifts battle zones between stagnation (Great House society) and anarchy (the drought-driven de-structuring of the 1100s AD). On the upside, the Edge of Chaos can be a time when a complex adaptive system can become spontaneous and alive. This is not what happened in Chaco Canyon.

Once the Great Houses closed their doors and ceased to distribute corn to the starving, the Chacoan world came undone, piece by piece. In Waldrop's own words, "The edge of chaos is where even the most entrenched old guard will eventually be overthrown" (1992). Indeed, Great House elite society had unraveled as its power continued to diminish and as the light of its once powerful aura dimmed, then extinguished.

Waldrop further asserts that, while a human organization operates on the Edge of Chaos, it often evicts the "old guard," replacing it with a calmer, more stable society (1992). And that is precisely what starving regional Native American communities attempted to do as the Chacoan world continued to fragment. The attempt failed as the Great Houses remained closed.

A Volatile, Thermodynamic Cultural State

About 1150 AD, Downtown Chaco Canyon apparently endeavored to reignite its aura by staging huge religious festivals at Chaco Canyon. But even a brief frenzy of expensive rituals received no response, rain or otherwise, from the heavens. This alone would have sent shockwaves through every tier of San Juan Basin society. Based on Nancy Akins's data as cited in Timothy Kohler and Kathryn Kramer Turner's article "Raiding for Women in the Pre-Hispanic Northern Pueblo Southwest?," conflicts between Chaco Canyon's small house populations continued to increase: including fighting, wife beating, starving infants, and the kinds of toxic anger that destroy any sense of order and defy cultural rules (Kohler and Turner 2006).

As noted above, by the early 1200s AD, many Great House families had already fled the San Juan Basin. In their wake, they left huge quantities of dried corn in their storerooms. Apparently there was no organized Great House effort to pass on these huge quantities of food to the starving small house corn farmers who had rendered up millions of tons of corn bounties over the course of three centuries.

Instead, caravans of Great House magnates and their sons fled north to the Montezuma Valley of Colorado. Some established miniature versions of Chaco Great House villages known as "scion communities." But based on the Montezuma Valley's local pottery styles, some archaeologists are convinced that these sons of Great House priests had abandoned their San Juan Basin families altogether when they moved north and married local Colorado women. A number of these scion settlements faded away after several generations.

Among the questions that archaeologists cannot answer is, What happened to the once carefully protected/sequestered Great House women back in Chaco Canyon? Why did they not accompany the young male scions to their new communities in Colorado? Were they left behind along with the stored corn and bowmen? If anyone has clues, they may be derived from the Diné "Beggar Woman's" tales published by noted ethnologist Klara B. Kelley and her Diné research partner Harris Francis (Kelley and Francis 2019).

One possibility is that previous decades of pellagra and its unpleasant side effect of uncontrollable diarrhea had tainted many hamlets' drinking water.

Lack of iron and protein in female farmers' diets generated risky pregnancies, generating a population in which men increasingly outnumbered women. A wide gender asymmetry probably contributed to the exit of young Great House males from the central San Juan Basin, who sought wives elsewhere.

According to multiple sources, still other caravans of Great House families headed south, following a magical/religious "meridian" (Lekson 1999). If Lekson's Chaco meridian theory is correct, exiting Great House caravans would have passed south through territories dominated by Zuni, Acoma, and Laguna pueblos—regional cultural powerhouses that had long supplied corn, firewood, animal hides, and the essential structural timbers needed to build and maintain the Great Houses. These districts had already begun to pull away from Chaco Canyon economy by the early 1100s, possibly unnerved by Chacoan pressure to cut more trees for Great House construction or to provide stored corn during droughts. In any case, the sight of large Great House entourages passing south would have signaled to anyone the dissolution of Chaco Canyon society.

Left behind in the wake of the caravans, roofs caved in on houses great and small, as there were no trees left to replace the beams. Cornfields sweltered under the relentless sun with fewer water girls and even fewer male laborers healthy enough to tend them. The upper canyon's water that had rolled downhill to Pueblo Bonito for five hundred years, harnessed by Chaco's elaborate wash system, had been reduced to a trickle.

Post Mortems on Great House Culture, 1130s–1210 AD

ECOLOGY

By the 1130s AD, the Chacoan Great House world had passed its ecological limits, overusing regional landscapes. Great House architecture had consumed several hundred thousand acres of prime, upland, protein-rich wild grass fields and denuded many thousands of acres of fir, spruce, cedar, piñón, and juniper. Great House–harvested roof beams had resulted in a loss of more than two hundred thousand mature trees. This deforestation in turn led to hotter regional air temperatures and increased water losses due to evaporation, and it wiped out piñón harvests—and the plants and animals that depended on them— and rendered firewood scarce. Those two hundred thousand trees have never grown back.

Poor environmental choices rapidly reduced plant and animal protein available to the general population in the San Juan Basin. The Chaco Canyon wash district went dry in the mid-1100s AD when a long drought persisted. Until then, water engineering had built a geographically expanding system—it had to be so, as Great House society consumed its own environment. Scholar John Holland's thoughts, cited in Waldrop (1992), come to mind here: "If I don't gather enough resources to make a copy of myself, I don't survive." The irony in this case is that Great Houses *had* gathered resources and stored huge quantities of food in ground-floor and sub-floor bunkers. And their kivas *were* paired. This was not accidental; two kivas were more robust than one.

The Great Houses *had* behaved precisely as John Holland's research suggested, but the doubling of stored food that might have saved thousands of small house families simply did not exist in most farmstead hamlets.

SOCIOLOGY

The world of ancient family sharing networks, which had once distributed food and information widely and reasonably, was displaced by a less generous world where food, information, and social power were restricted to the elite Great House lineages. Great House power was both commercial and magico-religious. Its effect was to further privilege the status of Great House ritual life. As the San Juan Basin's environment continued to degrade, the Great Houses' socioeconomic power rose. Slowly but steadily they co-opted the ancient sharing networks and, in the process, regained primacy of knowledge over regional crop and harvest patterns.

Complexity theory states that, over time, a society living on the Edge of Chaos tends to move agents (the Great House elites) in the direction of greater and greater complexity (Waldrop 1992). That trend, of course, leads to higher systemic costs and induces frenetic rates of cultural change. In the face of droughts of the mid-1100s AD, Great House life indeed became more complex, shifting trade patterns, investing in a class of bowmen as security forces to protect the elites and their stored food, and building watch towers

208 Chapter 13

to guard and control their roads. The Great House society's Edge of Chaos state demanded a fight to preserve their own perks. As the elites began to abandon the San Juan Basin, a huge regional social structure, five hundred years in the making, rapidly degraded into widespread chaos.

PSYCHOLOGY

By the 1200s AD, hunger and fear reigned in the central San Juan Basin. Violence, hopelessness, and a profound sense of helplessness combined to destroy local culture. It is possible many of the angrier people who were left abandoned by Great House elites suffered from pellagra-driven insanity. In such a state, decision-making became flawed, lives were shortened, violence increased, and families disintegrated.

One important psychological factor about the fall of Chaco Canyon has yet to be researched or explained: when the Great House priests and their entourage fled the San Juan Basin, abandoning their staggering quantities of

Photo 16. El Faro (the light house), looking west. The elevated kiva and signal tower were built next to the Great North Road in the early 1100s, when uneasy residents of Chaco's great houses walled their courtyards and built control gates where roadways passed village walls. Archaeologist Steve Lekson is on the pinnacle. (Courtesy NPS, Collection 0002/043.004, Negative 12758.)

Photo 17. Kutz Canyon at the terminus of the Great North Road from Chaco. Nearby, an ancient wooden staircase descends to the canyon floor. Archaeologist Michael Marshall argues that this sacred place represented an entry into the Chacoan underworld where souls departed this earth to await eventual rebirth. Tabular sandstone blocks, remains of a roadside shrine or way station, clutter the hillock (*center*). The archaeologist in right foreground is unidentified. (Courtesy NPS, Collection 0002/043.004, Negative 12491.)

dried corn, why was that remaining corn *never* consumed by the starving communities abandoned by the Great House elites, nor anyone else? Even today, there are unexcavated ruins of Great Houses, *still* protecting tons of ancient corn cobs.

THERMODYNAMICS

Great Houses were expensively built cultural memes: a multistoried Great House cost billions of work calories to create an abode that projected safety, order, and predictability. When that illusion was destroyed in the 1100s AD, Chacoan society exploded like a grenade. Small house families on the axis of entropy focused on both survival and revenge.

At huge costs, the Great Houses gathered energy, water, and mystical power, but they could not and would not refocus on the higher levels of food sharing with the general population when it was most needed. By the early 1200s AD, the Great House world had burnt out, leaving a few cinders of new Great Houses like Chimney Rock in Colorado and a rich legacy of women's gardening knowledge and techniques that have survived and are still practiced in today's Pueblo communities.

The Fate of Small House Farmers

As tragic and as chaotic as the 1150s to mid-1200s AD may have been for thousands of farmers, a number of families did manage to survive. Family groups sought safety in a growing number of communities in the extensive canyon country of the northern San Juan Basin. Others fled into the uplands built palisaded farmsteads atop rugged mesas, or dug deep pithouses surrounded by palisades in elevations typically above six thousand feet. Death rates in the region were probably staggering, as both violence and starvation took their toll.

By the 1300s AD, amid another series of droughts, new religions like the Kachina cults founded themselves on a more pragmatic set of cultural values, erasing much of the former Chacoan magico-religious patterns. Some groups of surviving Pueblo peoples reorganized into large, well-knit communities that were more complex in social structure and far more equally balanced in their agricultural goals than Chaco had been. Pueblo people speaking Towa, Tewa, and Tano languages began to populate the Upper Rio Grande district, where they live to this day.

Women's Reproductive Dynamics Turned Upside Down

An overlooked phenomenon regarding Chaco Canyon is that corn-based diets led to an epic demographic disaster. The fabled Great House era of the 900s–1100s AD, clever in so many ways, had turned the previous centuries' stable and successful dynamics of procreation upside down. Infant mortality spiked during this time, as did malnutrition among young mothers.

Discussed at length in this book are the women of the Late Archaic who foraged upland ricegrasses and became the first broad-spectrum focused gardeners in the 500s BC. These women consumed high amounts of protein thanks to those upland grass seeds and balanced their diets with a variety of seeds, nuts, leafy greens, berries, and *moderate* consumption of small-cobbed corn. They had sufficient protein in their diets that the time between births lengthened to thirty-two to thirty-four months, nearly *three* years. These slower birthrates had the effect of reducing infant and mother mortality, as women's bodies were allowed time to return to their physiological peak before another pregnancy.

These healthy women gardeners were the immediate antecedents of the peoples who would eventually become the stable and successful Basketmaker society. The advantage of their highly diverse gardens and diets became an efficient and "hidden" demographic factor (for archaeologists) as Late Basketmaker society moved into Chacoan times. It is probable that these women had been invited to live at Pueblo Bonito around the 800s AD, in order for the elites to reap the benefits of their good biology and childbearing successes. After all, these women represented a culture where cycles of birth, famine, and early death were all minimized.

Over time, however, the Chacoan world, perhaps blinded by hubris, misapprehended the dynamics and dietary differences that had made the Basketmaker women successful. When Chaco Canyon privileged corn as an apex food, that key diet-based demographic of the Basketmaker era dissolved. Women's lives thereafter consisted of corn-heavy diets, less ricegrass to forage, tainted household water, iron deficiency anemia, and the ugly ravages of pellagra. Add to that the long hours of grinding corn that tragically reshaped young women's pelvis girdles and the time span between pregnancies among small-house women in Chaco Canyon reduced to around twenty-five months, or roughly *two* years.

Shortening the pregnancy spacing required more daily protein per family. But in the droughts of the 1100s AD, there was no food margin to feed children. Closely placed pregnancies also generated high rates of infant mortality—over time, fewer and fewer children would be born into villages where healthy mothers could produce enough breast milk to properly support their infants.

By the early 1200s AD, the number of healthy young women had become a shrinking component of the San Juan Basin's regional population. The huge economic and psychological losses of countless failed pregnancies had completely erased the beneficial demographics Chaco had inherited from the ancient Basketmaker era. Alfred Lotka's law applies in this case: the cost of thousands upon thousands of failed pregnancies and infant mortality each year bankrupted many communities. There simply was no way to overcome the lack of clean water and dietary protein, healthy laborers, and psychic losses of infant deaths. And Chacoan society, like any other, was destined to rise—or fall—based on the well-being of its populations.

Butterfly Woman

The wisdom of ancient Puebloan society is clear: They understood the dangers of too much corn. This powerful lesson is embodied in a Zuni-style Kachina painting from the 1300s–1400s AD, depicting Butterfly Woman, a very sacred young spirit with a large, rosy, butterfly-shaped skin rash across her cheeks and forehead.[1] She, of course, shows the classic dermatological symptoms of early-stage pellagra (Etheridge 1972). Her associated spirit partner is known in English as the "Meat Eater," a male Kachina who flies through the late winter skies with her to make certain that families had stored enough dried turkey meat to feed it to their children in late winter and early spring.[2] Turkey meat, as noted, was the most effective protein antidote to pellagra. Raising turkeys had become common among later Chacoan era communities and has flourished in Pueblo communities until modern times. Turkeys not only diminished pellagra, they ate grasshoppers by the bushel, laid eggs, each of which provided a half-ounce of protein and fat to two young family members, and provided lush inner feathers woven into wondrous and ingenious thermal blankets.

Butterfly Woman's spirit is the proof of ancient Puebloans' sophisticated medical knowledge of a condition that would not be described nor labeled by the European world until two Italian doctors described pellagra in the 1790s AD, nearly six hundred years later.

The Dawn of a New World

By the 1300s AD, many Pueblo peoples continued to move east to remake

Photo 18. The mesa-top Hopi Pueblo of Shimopovi (also Shongopovi), Second Mesa, Arizona, about 1896. Note the kiva (center) and the ladder entryway-smokehole through its cribbed log roof, much like Ancestral Puebloan ones. The surrounding house blocks of adobe-plastered stone create a small plaza area (foreground). Corn and other necessities dry on the rafters. Except for the beams cut with steel axes and the plank doors (upper left and far right), this scene could have been photographed any time after 1000 AD. (Courtesy NPS, Collection 0028/006.002, Negative 77705. Callout on p. 201.)

their lives in the forested uplands of the upper Rio Grande district. Simultaneously, the Navajo peoples of Apache origin had begun to repopulate a number of abandoned Chacoan settlements in the dry, northern San Juan Basin (Brugge 2018). In short, new societies had emerged from the fragments of the Great House disaster.

One legacy of lessons learned when the Chacoan world failed was that Pueblo communities sought to incorporate village lands offering a variety of ecozones. Corn continued to be a primary crop, but virtually every community emphasized crop diversity in their gardens. Amaranth again thrived in new, carefully chosen garden plots. Many large Pueblo communities were surrounded by extensive patches of wolfberry, yucca, and iron-rich chenopod leaves. Juniper berries, acorn processing, and piñón nuts resurged. Because regional population had been dramatically reduced, grasslands flourished in the Rio Grande watershed.

Photo 19. House and Hopi family near plaza at Walpi on high, narrow First Mesa, Arizona, about 1896. The boy at right wears his hair in the traditional style. His deerskin moccasins are locally made but his Levis are manufactured. (Courtesy NPS, Collection 0028/006.002, Negative 77715.)

The pulsing star of energy, rhythmically blowing hot then cool throughout its life cycle and giving Chaco Canyon its aura and expanse, was supplanted in the ensuing centuries by something smaller, closer-knit, and more carefully defined: Puebloans chose lands that ranged from valley bottoms to mountain peaks or high mesas. Community access to a wide variety of ecotones was built into Puebloan social structure and practice. The Pueblo people spread across their landscape each spring, planting in diverse zones. With the fall harvest they returned to their Pueblos, which became "center place" (Stuart 2010).

Tightly organized Puebloan communities minimized conflict for centuries. Droughts of the late 1200s–1400s AD would again cast deep shadows on personal well-being, and eventually the Spanish would arrive, again turning the Puebloan world upside down.

In the meantime, however, life in the Rio Grande district had offered calm, order, and enough food to store for most winters. New cultural norms emerged. Juniper trees, piñón nuts, and wolfberries abounded, and high-protein ricegrasses earned their space in family storerooms once again. Large flocks of domesticated turkeys returned to cluck happily in the morning sun. And in gardens across the Rio Grande watersheds, women's pocket gardens also flourished, filled with their diverse and life-sustaining crops of corn, beans, squash, amaranth, chenopods, and other iron-rich leaves.

As every demographer knows, the long-term demographic fate of a human society depends on the health, fecundity, and survival of women and their children. Rapid birth spacings were very risky in the late 1100s and early 1200s. Poor diets led to pellagra, and pellagra led to measurable and horrifying infant and mother deaths. Societies that lose a generation of young men suffer tragedy, but societies that lose a generation of potential mothers dissolve. Chacoans were just like the Ona warriors on Tierra del Fuego Island, who hunted down women from other bands and brought their own ten-thousand-year-old society to extinction in just thirty years.

The portion of the Southwest now known as New Mexico includes many ancient, rich traditions. Descendants of those who once lived in the Chacoan world are their legacy, their survivors, and they are now our neighbors. Many still live traditional lives, but many also are also our nurses, college students, politicians, lawyers, teachers, our most advanced artists, and, best of all, our friends.

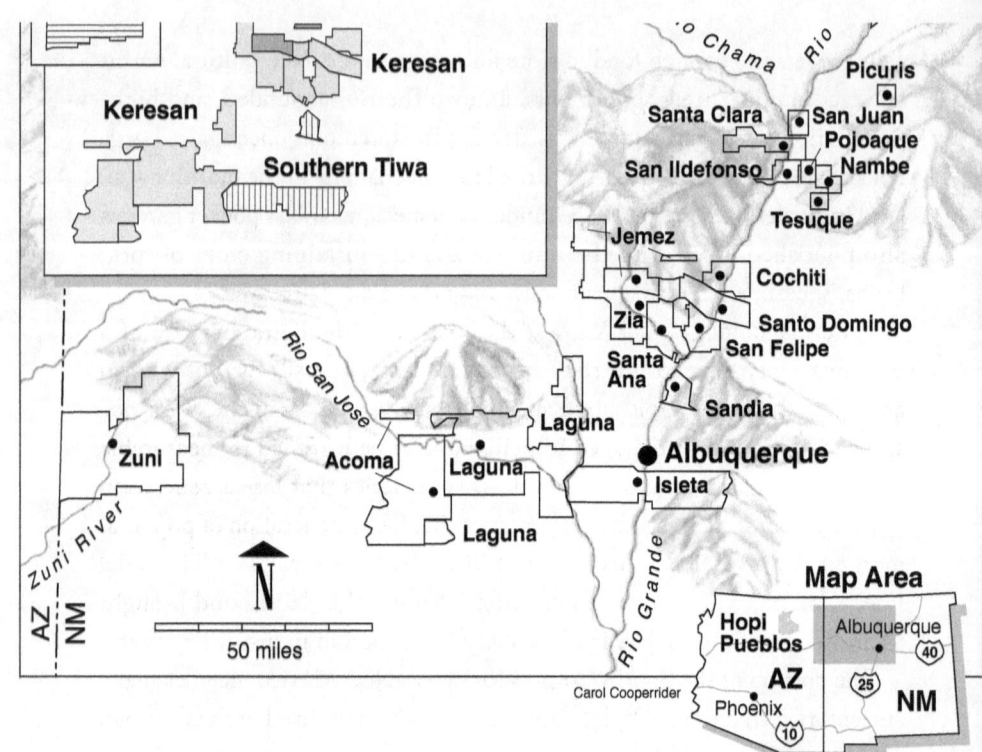

Map 5. The Pueblos of modern times. Map by Carol Cooper Rider.

EPILOGUE

PUEBLO WOMEN

Past, Present, and Future in Knowledge, in Change, in Balance

ELIZABETH AKIYA CHESTNUT

"Fieldwork requires both men and women to really get a whole picture of culture."

—RUTH BUNZEL, ANTHROPOLOGIST, ZUNI (1929)

THIS CODA TO DAVID Stuart's work is the result of an ongoing discussion about pre-European contact status and achievements of Pueblo women with Elizabeth Akiya Chestnut (History/Anthropology; Indian Pueblo Cultural Center Museum, past director). In the early part of this book, Stuart paints a vivid picture of the ancient female gardener/gatherers who by about 500 BC pioneered healthy, high-protein, broad spectrum-based diets. Women's high status remains evident among all Pueblos for the critical year-to-year decision-making of those who worked with various kinds of food sources and chose the ones that could withstand the rigors of an unforgiving, often unpredictable climate. This is perhaps most evident in the Western Pueblos (Hopi, Zuni, Acoma), where women still own the fields and houses and still wield considerable political and religious influence that has not always been acknowledged or understood in field ethnographies, past and present.

This epilogue is based in an ongoing discussion between David Stuart and this writer, focused on two Pueblo women achieving prominent national status. First, Deb Haaland (Laguna), a lawyer and successful New Mexico political organizer, was elected to the US House of Representatives, then nominated by President Biden and confirmed as the first woman, Native American secretary of the interior. Second, Cynthia Chavez Lamar (San Felipe/Hopi-Tewa/Navajo) negotiated higher education and professional challenges to be appointed as the first woman director of the Smithsonian National Museum of the American Indian (Washington, DC/New York sites). Such attainments by two Native, Pueblo women is unprecedented. In New Mexico, where nineteen Pueblo villages still exist, other Pueblo women have achieved similarly at local levels. The significant achievements and outstanding performances of many Pueblo women today may be attributed to the unique regional conditions that have enabled surviving Pueblo communities to remain in their thirteenth-century locations. This continuity has allowed these communities to maintain deeply rooted religious and cultural traditions. Sex roles have adapted and evolved with changing conditions to meet substantial challenges in physical, social, political, and environmental contexts.

In older archaeological analyses, Pueblo men were assumed to be the main source of the spectacular material remains of the vast Chacoan world. Much has been written about ancient pre–European contact Pueblo material remains, structurally complex stone architectural patterns, religious units (kivas), ritual artifacts, sophisticated water management systems, agricultural methods as well as domestic products including pottery, weaving, and other arts. Early twentieth-century ethnologists, some of them women, worked with, interacted with, and documented contemporary Pueblo communities. What female researchers learned suggested a more equal situation for Pueblo women than admitted to by male cohorts. As for ancient patterns, recent analysis of archaeological data, including DNA, from Chaco suggests that Pueblo women played significant roles in the lived culture and were revered, celebrated, and honored. Perhaps, as Stuart noted, this may date as far back as pre-sedentary agricultural times, when women, as foragers, might have begun retaining gathering rights for specific areas. After corn and sedentary agriculture became vital hallmarks of Pueblo existence, as Stuart noted, through the Chaco period (as is still the case at Hopi), women retained founder-use rights of specific fields for their families or clans and made critical planting decisions regarding appropriate seed-corn choices for a given year.

For both Pueblo past and present, this author suggests that for Pueblo people, the depth of beliefs, patterns of behavior and interaction, even its material manifestations may be understood from a group-centered "reciprocal obligation" perspective, whereby complementary interaction is foundational to society. Men and women both have roles to fulfill and tasks to be done, with lives turning on slim margins dictated by water scarcity. Thus, both Pueblo men and women subsume individual goals, extolled in contemporary Euro-American terms, to that of a group, an intergenerational family, a clan, one or more sodalities (non–family member group controlling banks of knowledge esoteric and/or practical) (Ware 2014), dual-sided village moiety membership, a village large or small.

Today, across linguistic, ethnic, and behavioral differences in the Pueblo world, the traditional roles of Pueblo women vary in detail and in the degree to which women participate in major decision-making with men. Yet "women have a definite place and expected behavior in Pueblo society. Their place is more equal than in our society" (Babcock and Parezo 1988, 125). As in older Chaco patterns, women in Western Pueblos (Hopi, Zuni, and Acoma) head clans, maintain house ownership, garden plots (Zuni), and may work fields and ranching units. Though men work land, Hopi women inherit agricultural land plots, lead clans, and organize major village-wide support for vital ceremonial functions (Anschuetz, in conversation with author, 2022). In addition to managing and guiding multigeneration family life, women may also belong to important, decision-wielding society (sodality) groups. Through such groups, Acoma women still control coveted old village housing assignments. Women also influence, in discussions with male groups, solutions to housing conundrums arising from contemporary non-marriage partnerships (Kurt Anscheutz, in conversation with author, 2022). At Hopi, a grandmother with deep, experiential, multigeneration knowledge has the power, on hearing field reports by male family members, to choose the appropriate seed corn type for a given planting season, wet or alarmingly dry (Kurt Anschuetz, in conversation with author, 2022). Even in Eastern Pueblos, among apparently male-dominant Tewa, land inheritance is bilateral, meaning that either a daughter or son is eligible to inherit after the death of a senior lineage holder, thus allowing women the possibility of land tenure and respect acknowledged (Dozier 1980; Ford 2014; Kennett 2017).

Without delving deeply into existing literature on this subject, there is also substantial researcher evidence of annual cycle rituals, dances, prayers, and songs of deep respect that pay symbolic homage to Pueblo reverence for women (Kurath/Garcia, 1970). First and foremost is the ubiquitous Corn Mother figure symbolizing the ultimate feminine power of procreation, central to life itself. There are also the Basket Dance and the Blue Corn Maiden (Tewa) dances. Similarly, Zuni Olla Maidens Dances celebrate the knowledge banks, strengths, flexibility, and persistence of Pueblo women under conditions of constant change. Women, symbolically, are basic, foundational to Pueblo life: They are life itself. At the Chaco Canyon Pueblo Bonito site, the graves of three women and their wealth of grave goods suggested a female lineage, high status, and reverence. DNA analysis—albeit without Pueblo permission—confirmed a lineage matriarchy, including a grandmother, mother, and daughter, perhaps of a founding family (Kennett 2017). The burial of a male identified as the brother of the grandmother indicated male roles. Then as now, men married out of a natal extended family but returned to teach knowledge arcane and practical, religion, and ritual to their sisters' male children. Done in kivas, this teaching was seen as a sign of male dominance. However, this ignores the complementary aspect of Pueblo sex roles, of women's knowledge and work input, without which the Pueblo world could not exist.

It may be impossible to definitively tie contemporary Pueblo women to deep, thousand-year agriculturally based patterns, but a wealth of ethnological literature and currently lived life patterns demonstrate that these ancient patterns, however modified, are alive in Pueblo villages. Though invaded, disrupted, dispersed, and subject to population crashes, regrouped Pueblos were never removed or exiled—as were most other Native groups—from their thirteenth-century occupation areas. Geographic isolation from high-desert barriers and cultural, legal, and (ironically) political buffers hindered Spanish plantation (encomienda)-style economies and helped to preserve Pueblo lifeways. Perhaps Pueblo beliefs regarding change as an assumed state, their recognition of the need to adapt and find new balance, a new center place, and their selectively adjusted knowledge bases retained with selective barriers, have favored their persistence into the present.

Economic forces beyond sedentary agriculture lured Pueblos toward cash income sources. With the arrival of the railroads, it was Pueblo women who took first advantage of the new situation, selling to tourists their anciently conceived and traded pottery for money income. Archaeologically vetted, these pots with complex painted designs were promoted by trading posts, but importantly by Santa Fe art museums both nationally and internationally. In the 1930s potter Maria Martinez (San Ildefonso), among others, became a household name. The success of such Pueblo women was not lost on other Pueblo women.

Links were also made between economic advantage and education. Pueblo people are very astute, sociological students of non-Pueblo people, whether they are Navajo, Apache, Spanish, Mexican, or American. The Pueblo people have long prioritized flexibility in what to adopt and what to ignore (Jojola, in conversation with author, 2018). Formal education, first basic, then higher, came to be seen as increasingly important in Bureau of Indian Affairs, Catholic, private, and then public institutions. More and more Pueblo women earned degrees in education and then taught in home communities. Graduate degrees, mainly for Pueblo men, increased slowly in the immediate post–World War II era. In the 1960s and '70s, federal legislation was passed specifically to support educational opportunities for women. The result was a 1970s explosion into higher education by Pueblo women in law and art. A small cohort of Pueblo women entered and graduated from University of New Mexico Law School. The Santa Fe Indian School promoted art education among Pueblo women; then the creation of the Santa Fe Institute for American Indian Arts further professionalized art studies. The rest is history.

Today, higher education among Pueblo women (and men) remains driven by Pueblo community needs: legal (land, water, business), scientific (biology, ecology, geology), historic (archaeology as history), art (techniques, management), architecture, entrepreneurship. Teaching at all its levels and manifestations also remains important for many Pueblo women, including applied linguistics for language retention. The point is that Pueblo women are enormously successful because they are supported by their families, cultural roots, and behavioral patterns grounded in meeting all challenges of survival daily and seasonally for many generations. Deb Haaland and Cynthia Chavez Lamar are the apogee, but many Pueblo women also benefit from the deep past and changing as context changes, many successfully. Deeper study of this subject from both the Pueblo community as well as those on the outside is warranted.

Pueblo acknowledgements: Pueblo women, their families, friends, and colleagues, Ohkay Owingeh/San Juan, Jemez, Acoma, Santa Clara, San Ildefonso, Santo Domingo, San Felipe, Zuni, Hopi, and the Indian Pueblo Cultural Center in Albuquerque, New Mexico. Individuals include, above all, the late Dr. Alfonso Ortiz (Oke Owingeh/San Juan) and Geronima Cruz Montoya, her sisters, and their daughters (Oke Owingeh Pueblo); the late Rina Swentzell (Santa Clara), writer of Pueblo women's understanding of Pueblo space/time, and her daughter Roxanne Swentzell, artist and sculptor; former governor and historian of Cochiti Pueblo Dr. Joseph Henry Suina; archivist at the Indian Pueblo Cultural Center Jonna Pedra (Acoma/Laguna); archivist of the Museum of Indian Arts and Culture Diane Bird (Santo Domingo/Cochiti); Penny Bird (Santo Domingo/Taos) of the New Mexico Public Education Department; Dr. Ted Jojola (Isleta) in the University of New Mexico Department of Architecture; and former Acoma governor and historic restoration professional Bryan Vallo, among many others. Archaeologists, yes, they are mostly male, but among the few who were truly sensitive to Pueblo women, their roles in survival and in critical decision-making past and present are Kurt Anschuetz, Richard Ford, and David E. Stuart. Finally, to late archaeologist Linda Cordell and those women writers whose work I've read in print but whom I've never had the opportunity to meet and discuss the issues: "Kunda wo hah" (Tewa, "thank you").—Elizabeth Akiya Chestnut, Albuquerque, New Mexico

Postscript

As I write this, virtually every university think tank and worldwide news network is telling us that scholars are shocked at the rapid decline in the vitality of our global environments. I am reminded that cultures are humans' complex adaptive systems. They operate under the same primordial rules that once slowly transformed matter into living things: energy, diversity, animal and human species, trees, plants, weather, climate, and the rest. There is one radical difference—humans have the capacity to imagine, invent, foresee the future, and express empathy.

Perhaps the complex system we call "The Industrial World" is now clinging to another Edge of Chaos, in which environment will be acknowledged as our most valuable asset and billionaires will become the ones to suddenly lose power. In other words, the sands of fate are shifting daily, and our "old guard" might fragment and fade away. History does not repeat itself, but natural thermodynamic processes do.

Notes

Introduction

1. For more genetic details, I recommend Jennifer Raff's *Origin: A Genetic History of the Americas* (Raff 2022), a book that lays it all out in understandable language.
2. A Navajo name in the Diné language for the ancient people who lived in the region before their arrival in the late 1300s or early 1400s AD.
3. Seaside and riverside populations subsisting on salmon operated under quite different ecological dynamics.
4. Newly introduced European diseases and later genocidal episodes contributed to shrinking Indigenous Australian populations.
5. Also knowns as "Yamana."
6. We watch the nightly news; the Southwest's ancient inhabitants watched nature. Most moderns are not trained to see the seasons, the landscapes, and to know their intricate, interacting parts. On our nightly news, a hurricane or tornado is "news." We no longer understand its underlying message of climatological dangers to come.
7. For a good analysis of cultural behaviors of chaos and complexity, I recommend James Gleick's *Chaos: Making a New Science* (1987).
8. I was first trained in archaeology while studying in Mexico City as an undergraduate student. Before that, I was a premed student in the United States who worked the night shift for several years in a large hospital emergency room.
9. Horticulture is human-initiated planting or enhancing of native plant food species. Ancient people's practice of selecting strains of amaranth or corn that produced the most seeds/kernels in fall and systematically carrying water to the plants qualifies as horticulture.

Chapter 1

1. This is how they are identified by climatologists and geologists.
2. Archaeologists have debated the Clovis cultural adaptations and precise dates for the last fifty years. They may still be debating another fifty years from now. Let's just accept the idea that it was a widespread and successful human adaptation in which important innovations to hunting techniques and foraging supported a fairly rapid rise

in population across a wide swath of the Southwest, which later spread rapidly to the southeastern United States.

3. Projectile boards used to launch long, thick arrows.

4. Between 1790 and 2022, 232 years after the United States gained efficiency by appropriating Native American lands and eliminating taxes and some import costs from Great Britain, it complexified, grew one-hundred-fold in population, and now lives in tense political times over its core cultural values. As a nation, we do not seem to understand that the greater a power phase, the more quickly it burns out.

5. The small human footprints, which have been described as those of tiny adults, are more likely children's footprints, made as they foraged pond plants near the drying lake's margins in the more recent climate period known as the Early Holocene.

6. Women's lifespans would have been on the low end of the spectrum due to childbirth risks.

7. Gomphothere hunts have been documented in southeastern Arizona by Paleo-Indian specialist Professor Bruce Huckell of the University of New Mexico.

8. See Sanchez et al. (2014) for a remarkably clear and well-illustrated article on the age and geographic range of Clovis.

9. Atole is an ancient mix of water and ground grains still common in the US Southwest and in rural Mexico. If local water is a bit alkaline/bitter, the atole grains softens the taste of "hard" water.

Chapter 2

1. Knowledge of the true number of Early and Early-Middle Archaic sites is shaped by their surface visibility on archaeological surveys. Many are buried three to six feet below today's ground level, especially in low-lying vales or locales where ancient topsoils have eroded away.

2. Those dugouts were stone-lined pits, to which hungry families might return. A few have been found unopened after two or three thousand years.

3. Most archaeologists refer to these as "pithouses"—a catchall descriptor. Changes in the size, depth, and architecture of pithouses were important from an adaptive evolutionary perspective.

4. This "award" system led to a few big-game hunters' Y-DNA predominating in their descending lineages.

5. This meant that the deleterious genes of high-status males containing autoimmune diseases or heart failure were disproportionately passed on to future generations.

6. That season, our smallish team was led by Dr. Joanne Townsend from the University of Manitoba. Our goal was to find traces of an ancient flake tool tradition, now variously called the Denbigh small-flake tradition or complex (Raff 2022).

7. Male DNA is divided at conception. Y-DNA is male. In contrast, female DNA is not divided at conception, as it is preserved intact in female eggs. It is known as mtDNA (mitochondrial). Thus, geneticists can trace all of our female lineages back to the origins of our species. Not so for the males.

8. Demographers have long known that male births in large populations generally exceed female by 1 to 3 percent. Yet young males die at greater rates than their female cohorts, as they engage in riskier behaviors.
9. Insane asylums in the United States were heavily populated by pellagra sufferers into the late 1920s.
10. The area around Zuni Pueblo in northwestern New Mexico may be another area where corn was grown experimentally at an early date. The seeds of that first small-cobbed corn most likely came from northwestern Mexico, where it had been partially domesticated for thousands of years (Gibbons 2022).

Chapter 3

1. The calculations of a population's predicted height, weight, and longevity are very complicated. A plain statement is that a single birth, a well-fed mom, and no great eras of famine can contribute to greater height and longevity potential as much as can genes. Twin births, eras of food scarcity, and malnourished mothers can contribute to shorter stature and somewhat shorter lives.
2. Horticulture is defined here as small-scale gardening and an experimental, part-time effort until major gains in crop selection and improvements outpace the food returns obtained by foraging for wild plants.
3. The world's largest volcanic bowl.
4. Many ethnographic examples of such cooperating groups are documented in the groundbreaking 1968 volumes of *Man the Hunter* (Lee 1968) and the subsequent special "hunter-gatherer" academic sessions supported by the American Anthropological Association.
5. Heavy labor demands on young women's protein-limited bodies delayed the onset of menses, just as it did in the eighteenth and nineteenth centuries in the United States.
6. Clusters of small camps with shallow dugout shelters have been found near bison kills from the era of roughly 2000 BC, signaling multiple contemporaneous lifeways.
7. Cottontail rabbit was high on the list of favored protein sources. Widespread evidence of cottontail leg bones in campsites tell us that spit-roasted rabbit haunches were *the* thing when available.

Chapter 4

1. A corn variety cannot be fully identified by its kernel's color and outward appearance. Even now, laboratory identification of corn variety depends heavily on micromeasurements of pollen circumference and mineral analysis.
2. At this point in history, we do not know if the lineage names attached to "first and exclusive foraging rights" were male or female progenitors. My assumption is female, based on the long Chaco-era and later Pueblo people's tradition of female land and house ownership.
3. The underlying clay deposits sealed the loss of surface moisture in two to four feet

of sandy soil. The Hopi people of western New Mexico, Arizona, and Utah became masters of corn cropping in sandy areas. Traditional Hopi families still plant their sand dune fields.

4. Cushaws are varieties of squashes. Several drought-resistant varieties likely made it to the San Juan Basin through Arizona and northern Mexico.

5. Many Ecuadorian peasant households held some common lands in what was left of their ancient community property. In the district I studied, most former Indigenous lands had wound up being owned by the hacienda class. A quarter-acre Indigenous woman's garden was typical. A few of the oldest families had the equivalent of a half-acre of garden land.

6. A reminder that complexity is the enemy of efficiency.

7. Larger dugout storage pits with sealed stone slab lids and disguising adobe mud plastered over all at ground level were harder to find by raiders and much more tedious to open.

8. Linguistic variation in the San Juan Basin during Chacoan times would become a distinguishing characteristic within a widespread culture that would later contribute to modern Pueblo society.

Chapter 5

1. Archaeologists are prone to think in named, cultural eras where tools, pottery, habitations, and food production are distinctive. Yet before a distinct era emerges, confusing overlap of behaviors and technologies are the norm. This is why named eras of culture change are rarely obvious and smooth.

2. Citing those rights likely involved some version of cultured memory memes reciting the women forebears by name, in order.

3. An ordinary garden size was about two thousand to three thousand square feet. A very large garden of this era was about a quarter acre.

4. Piñón trees do not produce nuts every year. Their cycle of production is generally three or four years between harvests.

Chapter 6

1. A lobed pithouse plan included a central circle with one or two bulging wings. Much of the extra space was used for storage.

2. About 90 percent of corn's protein resides in its tassels—*not* in the cob.

3. Wolfberries could be used as a poultice, but most were dried and eaten alone or added to meals. They offered fiber, sugar, and nutritional compounds that protected the eyes from the intense ultraviolet rays of high country occupation, essential to ancient sky watchers. They are cousins of China's goji berries.

4. In southern Ecuador of 1969–1970, I logged lots of rough data about bathroom practices, most of which took place between the corn or sugarcane rows in Quichua-speaking subsistence farmers' house gardens. Even the donkeys were occasionally paraded through

family fields of quinoa, potatoes, sunflowers, and corn plots used for chicken feed. Chickens and *cuy* (guinea pigs) did their share of soil enrichment as well.
5. Cut or burned to size. Cutting timber with stone axes is quite like cutting firewood to length with a dull bladed splitting maul.
6. *Not* a Hispanic invention, despite claims by some regional scholars.

Chapter 7

1. Turkey and turkey feather remains found in the very Late Archaic period date to about 100–200 BC in south-central and western New Mexico, from a time when most families still foraged for a living, even though seeds of corn, beans, and squash were all available (McBrinn 2018).
2. "Rio San Jose": Still owned and farmed by the people of Acoma Pueblo. The Rio Puerco got its name by Hispanic settlers in the 1600s; thus, it is not an ancient name. "Puerco" suggests pigs, who root in the soggy, turgid mud of creek bottoms.
3. The Appalachians included Scots, Scots-Irish, German, English, Hessian soldiers, Black "maroons," and many people with mixed Native American ancestry.
4. Some were camouflaged by adobe clay as a bland patch of uninteresting hillside.
5. The district from the Arizona state line to Gallup and the Puerco River of the west.
6. This number is based partly on hearth counts.

Chapter 8

1. In modern New Mexico, many gardens are still cobbled or brick-edged. Plant a few flowers along that row and you have created a tiny ecotone adjacent to your yard grass. That thirty-foot-long ecotone will draw birds, small animals, butterflies, resident skunks, or even coyotes.
2. Throughout the Americas, many wash districts have been found in dry regions. All of them involve gravity, acequias, and local populations who clean and maintain the waterways and participate in seasonal rituals that are integral to acequia culture.

Chapter 9

1. Corn pollen is plentiful on the silk under the outer husk layers; however, it is scarce in cobs stripped of the coarse outer husk.
2. In 1992 noted field archaeologist John R. Stein and several of his colleagues from the Navajo Nation's Cultural Resource Division presented a stunning account in my "Ancient New Mexico" class at the University of New Mexico of the growth of Chacoan roads during the 800s–900s AD, which were based on earlier trails converted during this era.
3. The numbers of increased individuals in the southern San Juan Basin need not have been huge to unbalance the caloric scale and create another human food dilemma.

Chapter 10

1. The descendants of those Pueblo peoples still produce valuable *heishe* necklaces sold in Santa Fe stores and on the portal of the Palace of the Governors fronting the plaza.
2. Were those males the women's fathers, sons, uncles, or spouses? Could a male small house commoner become a Great House priest, or was that role restricted to those males borne by the ranking female lineage at Pueblo Bonito?
3. As a university student in Mexico City in 1965–1966, I witnessed the excavation of such jars at Pueblo Perdido, a rich and long-lived village a few miles north of Mexico City.
4. In fact, some archaeologists do not define a Chacoan style structure as a "Great House" unless a paired kiva is present.
5. My label of "outpost Great Houses" does not follow the labeling of most archaeologists. I think it unlikely that each one of those outposts housed elite female landowners genetically connected to Pueblo Bonito.
6. Human urges to procreate are enhanced by disasters like famines, wars, plagues, and high infant mortality.
7. The village is also known by its archaeological record name, site 29SJ1659. This code indicates federal site number SJ for San Juan County, New Mexico, the 1659th site recorded. Site cards for Shabik'eshchee are filed in both Santa Fe's ARMS files and federal databases.
8. My own tentative answer is that it was ancient female lineage land, linked to earlier styles of agriculture, architecture, and farming techniques. It radiates a cultural memory of early Basketmaker origins.
9. Hopi are still residents of northeastern Arizona's mesa and wash country.

Chapter 11

1. "Tall house" in the Navajo/Diné language.
2. Linguistic echoes from Chaco's past still persist in contemporary Pueblo society, where they are kept private in kiva societies and in sacred narrative and prayer.
3. A quipu is an ancient South American accounting device of knots made in animal or plant fibers.
4. It can be argued that four to six construction surges took place. I imagine a young man whose mother's or aunt's corn plots had failed leaving home and joining a Great House road construction crew.

Chapter 12

1. Think of modern American water rights, and just how consistently the water and land rights of Native Americans have been ignored in spite of presidentially signed treaties.

Chapter 13

1. Also spelled Kat'sina.
2. This is a rough English translation, and an Anglo interpretation of her identity.

Bibliography

Adams, K. R., C. M. Meegan, S. G. Ortman, et al. 2006. "MAÍS (Maize of American Indigenous Societies) Southwest: Ear Descriptions and Traits that Distinguish 27 Morphologically Distinct Groups of 123 Historic USDA Maize (Zea mays L. ssp. mays)." *Accessions, and Data Relevant to Archaeological Subsistence Models.* Arizona State University.

Akins, Nancy J. 1986. *A Biocultural Approach to Human Burials from Chaco Canyon, New Mexico.* Reports of the Chaco Center no. 9. National Park Service, US Department of the Interior.

Anscheutz, Kurt. 2010. "Women are Corn, Men are Rain: Agriculture and Movement Among the Tewa in North Central New Mexico between A.D. 1250 and 1598." *Papers of the Archaeological Society of New Mexico* 36.

Babcock, Barbara A. and Nancy J. Parezo. 1988. "Florence Hawley Ellis." In *Daughters of the Desert Women Anthropologists and the Native American Southwest, 1880–1980* by Barbara A. Babcock and Nancy J. Parezo. University of New Mexico Press.

Battillo, J. 2018. "The Role of Corn Fungus in Basketmaker II Diet: A Paleonutrition Perspective on Early Corn Farming Adaptations." *Journal of Archaeological Science* 21:64–69.

Bennison, Sarah. 2023. *The Entablo Manuscript: Water Rituals and Khipu Boards of San Pedro de Casta, Peru.* University of Texas Press.

Bernardini, Wesley. 1999. "Reassessing the Scale of Social Interaction at Pueblo Bonito, Chaco Canyon, New Mexico." *Kivas* 64 (4): 447–70.

Betancourt, J. L., T. R. Vandevender, and P. S. Martin. 1990. *Packrat Middens: The Last 40,000 Years of Biotic Change.* University of Arizona Press.

Bridges, E. Lucas. 1948. *Uttermost Part of the Earth.* London: Hodder & Stoughton.

Brugge, David M. 2018. *A History of the Chaco Navajos.* Forgotten Books.

Bunzel, Ruth. 1929. *The Pueblo Potter: A Study in Creative Imagination in Primitive Art.* Contributions to American Ethnology. New York: Columbia University Press.

Burns, B. T. 1983. "Simulated Anasazi Storage Behavior Using Crop Yields Reconstructed from Tree-Rings: A.D. 652–1968." PhD diss., University of Arizona.

Carpenter, John, Guadalupe Sanchez, and Ismael Sanchez. 2018. "The Archaic Period in Sonora." In Vierra, *Archaic Southwest.*

Convert To. "Corn Grains Kernels Amounts Converter." http://convert-to.com/507/yellow-dry-corn-grain-kernels-amounts-conversion.html.

Crown, Patricia L. 2000a. "Women's Role in Changing Cuisine." In Crown, *Women and Men in the Prehispanic Southwest.*

Crown, Patricia L. 2000b. *Women and Men in the Prehispanic Southwest.* School of American Research.

Crown, Patricia L., and Suzanne K. Fish. 1996. "Gender and Status in the Hohokam Pre-Classic to Classic Transition." *American Anthropologist* 98 (4): 803–17.

Dillehay, Thomas D. 1997. *Monte Verde: A Late Pleistocene Settlement in Chile.* Smithsonian Institution Press.

Dixon, E. James. 1999. *Bones, Boats, and Bison.* University of New Mexico Press.

Doyel, D.E., D. Breternitz, and M. P. Marshall. 1984. "Reports of the Chaco Center." National Park Service, Albuquerque, 37–54.

Dozier, Edward P. 1983. *The Pueblo Indians of North America.* Waveland Press.

Dunmire, W., and Gail D. Tierney. 1997. *Wild Plants and Native Peoples of the Four Corners.* Museum of New Mexico Press.

Dunmire, William W., and Gail D. Tierney. 1995. *Wild Plants of the Pueblo Province.* Museum of New Mexico Press.

Etheridge, Elizabeth. 1972. *The Butterfly Class: A Social History of Pellagra in the South.* Greenwood.

Fogel, Robert W. 2012. *Explaining Long-Term Trends in Health and Longevity.* Cambridge University Press.

Ford, Richard. 2018. "Matu'in: The Bridge Between Kinship and Clan in the Tewa Pueblos of New Mexico." In *Puebloan Societies: Homology and Heterogeneity in Time and Space*, ed. Peter M. Whiteley. University of New Mexico Press.

Friedman, Richard, John R. Stein, and Taft Blackhorse. 2003. "A Study of Pre-Columbian Irrigation System at Newcomb, New Mexico." *Journal of GIS Archaeology* 1:1–10.

Gibbon, Edward. 1996. *The History of the Decline and Fall of the Roman Empire.* Edited by David Womersley. Penguin Classics.

Gibbons, Ann. 2022. "Southern Roots for the Maya—and the Maize that Fed Them." *Science* 375 (6587): 1325.

Gillman, Joseph, and Theodore Gillman. 1951. *Perspectives in Human Malnutrition.* Grune and Stratton.

Gleick, James. 1987. *Chaos: Making a New Science.* Viking.

Hall, Stephen A. 2018. "Paleoenvironments of the American Southwest." In Vierra, *Archaic Southwest.*

Hanselka, J. Kevin. 2018. "A Pan-Regional Overview of Archaic Agriculture in the Southwest." In Vierra, *Archaic Southwest.*

Heitman, Carrie C., and Phil R. Geib. 2015. "The Relevance of Maize Pollen for Assessing the Extent of Maize Production in Chaco Canyon." In Hetiman and Heitman, *Chaco Revisited.*

Heitman, Carrie C., and Stephen Plog Heitman, eds. 2015. *Chaco Revisited: New Research on the Prehistory of Chaco Canyon, New Mexico.* University of Arizona Press.

Jolie, Edward, and Laurie D. Webster. 2015. "A Perishable Perspective on Chacoan Social Identities." In Heitman and Heitman, *Chaco Revisited.*

Kantner, John. 2023. "Chaco Roads." Archaeology Southwest. https://www.archaeologysouthwest.org/wp-content/uploads/Chaco-Roads.pdf.

Kearns, Timothy. 2018. "Archaic Time and Distance in the San Juan Basin." In Vierra, *Archaic Southwest.*

Kelley, Klara B., and Harris Francis. 2019. *A Diné History of Navajoland.* University of Arizona Press.

Kennett, Douglas J., Stephen Plog, Richard J. George, et al. 2017. "Archaeogenomic evidence reveals prehistoric matrilineal dynasty." *Nature Communications* 8, (14115). https://doi.org/10.1038/ncomms14115.

Kimmerer, Robin Wall. 2015. *Braiding Sweetgrass: Indigenous Wisdom, Scientific Knowledge, and the Teachings of Plants.* Milkweed Editions.

Kohler, Timothy A., and Kathryn Kramer Turner. 2006. "Raiding for Women in the Pre-Hispanic Northern Pueblo Southwest?: A Pilot Examination." *Current Anthropology* 47 (6): 1035–45.

Kurath, Gertrude Prokosch, and Anthonio Garcia. 1970. *Music and Dance of the Tewa Pueblos.* Museum of New Mexico Press.

Lee, Richard B., and Irven DeVore. 1968. *Man the Hunter.* Transaction Publishers.

Leffer, Lauren. 2024. "Making Babies May Take 10 Times More Energy than We Thought." *Popular Science* online, May 16, 2024.

Lekson, Stephen H. 1999. *The Chaco Meridian.* Altamira Press.

Lekson, Stephen H. 2006. "Lords of the Great House: Pueblo Bonito as a Palace." In *Palace and Power in the Americas: From Peru to the Northwest Coast,* edited by Jessica Joyce and Patricia Joan Sarro Christie. University of Texas Press.

Lotka, Alfred J. 1922. "Contribution to the Energetics of Evolution." *Proceedings of the National Academy of Sciences of the United States of America* 8 (6): 147–51.

MacWilliams, A. C. 2018. "Archaic Chihuahua: Many Points, Few Sites." In Vierra, *Archaic Southwest.*

Marden, Kerriann. 2015. "Human Burials of Chaco Canyon: New Developments in Cultural Interpretations Through Skeletal Analysis." In Heitman and Heitman, *Chaco Revisited.*

Marshall, Michael P., John R. Stein, Richard W. Loose, and Judith E. Novotny. 1979. *Anasazi Communities of the San Juan Basin.* Public Service Company of New Mexico.

Martin, Debra. 2000. "Bodies and Lives: Biological Indicators of Health Differentials and Division of Labor by Sex." In Crown, *Women and Men in the Prehispanic Southwest.*

Martin, Paul A., and H. E. Wright. 1967. *Pleistocene Extinctions.* Proceedings of the VII Congress of the International Association for Quaternary Research. Yale Univeristy Press.

Mathien, Frances J., ed. 2010. *The Casamero Community in the Red Mesa Valley of Northwestern New Mexico.* New Mexico Bureau of Land Management Cultural Research Series. Archaeological Society of New Mexico.

McBrinn, Maxine E. 2018. "Foragers and Early Forager/Farmers in the Mogollon Highlands." In Vierra, *Archaic Southwest*.

Mills, Barbara J. 2015. "Unpacking the House: Ritual Practice and Social Networks at Chaco." In Heitman and Heitman, *Chaco Revisited*.

Minnis, Paul E. 2015. "Looking North Toward Chaco with Awe and Envy . . . Mostly." In Heitman and Heitman, *Chaco Revisited*.

Montgomery, John L. 2018. "The Archaic of Eastern New Mexico." In Vierra, *Archaic Southwest*.

Oyle, D., L. St. John, and T. Jones. 2013. *Plant Guide for Indian Ricegrass (Achnatherum oryzopsis)*. Aberdeen Plant Materials Center, Aberdeen: USDA-Natural Resources Division, Conservation Service.

Potter, Ben A., Joel D. Irish, Joshua D. Reuther, Carol Gelvin-Reymiller, and Vance T. Holliday. 2011. "A Terminal Pleistocene Child Cremation and Residential Structure from Eastern Beringia." *Science* 331 (6020): 1058–62.

Raff, Jennifer. 2022. *Origin: A Genetic History of the Americas*. Twelve Publishing.

Reed, Paul. 2011. Basketmaker III Synthesis. In *MAPL Project Final Report*. Woods Canyon Archaeological Consultants, Inc.

Reed, Paul F., and Phil R. Geib. 2013. "Sedentism, Social Change, Warfare and the Bow and Arrow in the Ancient Pueblo Southwest." *Journal of Evolutionary Anthropology* 22: 103–10.

Reinhard, Karl. 2000. "Coprolite Analysis: The Analysis of Ancient Human Feces for Dietary Data." In *Archaeological Method and Theory: An Encyclopedia*, edited by Linda Ellis. Garland.

Roe, Daphne A., MD. 1973. *A Plague of Corn: The Social History of Pellagra*. Cornell University Press.

Roth, Barbara. 2018. "The Southwestern Archaic: Divergent and Convergent Perspectives." In Vierra, *Archaic Southwest*.

Salfisburg, C., R. Cordero, and R. Dello-Russo. 2014. "The Hilltop Bison Site." *News-MAC: Newsletter of the New Mexico Archaeological Council*, no. 3, 8–12.

Sanchez, Guadalupe, Vance T. Holliday, Edmund P. Gaines, et al. 2014. "Human (Clovis)–gomphothere (Cuvieronius sp.) association ~13,390 calibrated yBP in Sonora, Mexico." *Proceedings of the National Academy of Sciences of the United States of America* 111, (30): 10,972–77.

Steckel, Richard. 2002. "A History of the Standard of Living in the United States." *Economic History Association Encyclopedia*, edited by Robert Whaples. Available at https://eh.net/encyclopedia/a-history-of-the-standard-of-living-in-the-united-states/.

Stiger, Mark. 2018. "Early Hunter-Gatherer Adaptations in the Upper Gunnison Basin of the Southern Rockies." In Vierra, *Archaic Southwest*.

Stuart, David E. 1972. "Band Structure and Ecological Variability: The Ona and Yahgan of Tierra del Fuego." PhD diss., University of New Mexico.

Stuart, David E. 1980. "Kinship and Social Organization in Tierra del Fuego:

Evolutionary Consquences." In *The Versatility of Kinship*, edited by Linda S. Cordell and S. J. Beckerman. Academic Press.

Stuart, David E. 2010. *Pueblo Peoples on the Pajarito Plateau: Archaeology and Efficiency*. University of New Mexico Press.

Stuart, David E. 2012. "Finding the Calories to Fuel the Chacoan World." School for Advanced Research, international lecture, YouTube video, September.

Stuart, David E. 2014. *Anasazi America*. 2nd ed. University of New Mexico Press.

Stuart, David E. 2019. *A Fragile Legacy of Well-Being: Three Families and the Trajectory of America, 1750–2019*. Cultural Impacts Publishing.

Sykes, Bryan. 2001. *The Seven Daughters of Eve: The Science that Reveals Our Genetic Ancestry*. Bantam Press.

Vierra, Bradley J. 1994. "Archaic Hunter-Gatherer Archaeology in the American Southwest." *ENMU Contributions in Anthropology* 13, no. 1.

Vierra, Bradley J., ed. 2018. *The Archaic Southwest: Foragers in an Arid Land*. University of Utah Press.

Vierra, Bradley J., and Richard I. Ford. 2006. "Early Maize Agriculture in the Northern Rio Grande Valley, New Mexico." In *Histories of Maize: Multidisciplinary Approaches to the Prehistory, Linguistics, Biogeography, Domestication, and Evolution of Maize*, edited by John E. Staller, Robert H. Tykot, and Bruce F. Benz. Routledge.

Vint, James, and Bradley J. Vierra. 2022. "Climate Change and the Neolithic in the American Southwest." *Journal of Anthropological Research* 78 (1).

Vivian, R. Gwinn, and Adam S. Watson. 2015. "Reevaluating and Modeling Agricultural Potential in the Chaco Core." In *Chaco Revisited: New Research on the Prehistory of Chaco Canyon, New Mexico*, edited by Carrie C. Heitman. University of Arizona Press.

Waldrop, M. Mitchell. 1992. *Complexity*. Simon & Schuster.

Ware, John. 2014. *A Pueblo Social History: Kinship, Sodality, and Community in the Northern Southwest*. School for Advanced Research Press.

Whittaker, J. C. 2012. "Ambiguous Endurance: Late Atl-Atls in the American Southwest?" *Kiva: Journal of Southwestern Anthropology and History* 78 (1): 79–98.

Windes, Thomas C. 2003. "This Old House: Construction and Abandonment at Pueblo Bonito." In *Pueblo Bonito: Center of the Chacoan World*, edited by Jill E. Neitzel. Smithsonian Books.

Index

Acoma, 206, 216–17, 219, 222; opening San Juan Basin to gardening, 59–60; and Rio San Jose, 229n2; status quo in ten and eleven hundreds AD, 197; traditional dancers at, 198–99
acorns, 6, 33, 38, 41, 53–54, 94; food efficiencies of foragers, 48–49; gardening efficiencies, 103; processing, 50, 57
ak-chin, 170
Akins, Nancy, 114, 132, 154, 186, 205, 230n4
Alaska, 2, 15, 19–20, 44
Altithermal Period: changing landscape in, 30–32; studying forager campsites, 47; territorial tensions, 33–34
amaranth, 2, 38–40, 69, 82, 96, 102, 108; flour, 156; and horticulture, 75–76; increasing growth of, 57; long path toward sustainable gardens, 85, 89; and rapid cultural changes, 65; seeds, 57, 156, 164; selecting, 225n9; and shift to gardening, 71; studying forager campsites, 47
American bison (*Bison bison*): and changing landscape in Alithermal period, 30–32; hunting in Paleo-Indian Southwest, 21–24
Anasazi America (Stuart), 88, 183–84, 190
Ancient Indians. *See* Paleo-Indians
animals, interactions with: bison, 21–24; dogs, 44; elephants, 29–30; turkey, 49–50
Animas River, 150, 163, 197

apogee (of Pueblo Bonito), period of, 194–96
Archaic Period. *See* Early Archaic Period, changes during; Early Middle Archaic Period; Late Archaic Period; Middle Archaic Period; Terminal Archaic Period
Archaic Southwest, The (Vierra), 88
architecture: complexity of, 123–24; compression of styles of, 122–23; traditional dating of specific periods, 121
Arizona. *See* San Juan Basin
assemblage, 56
atl-atls, 18, 21, 24, 55, 64, 70, 94, 154; eclipsing, 59; replacing, 101, 104–5; and social dynamics of hunters, 43–44; and territorial tensions, 33–34
atole, 33, 226n9
avenues to power, modifying landscapes as: cornfields reshaping daily life, 183–86; Great House expansion and grandeur, 179–81; overview, 177–78; powerful sense of "belonging," 186–87; Pueblo Bonito grandeur, 181–83; risk vs. reward in corn economy, 187–88; social complexities of Great House elites, 178–79

"Band Structure and Ecological Variability: The Ona and Yahgan of Tierra del Fuego," 57–58

Basketmaker era, 88; gardening expansion during, 105–8; gardening progress in, 81–82; identifying Basketmaker II period within, 101–3; shift to horticulture of, 80–81; steps in long path to sustainable gardens, 84–85

Basketmaker II era: complexity increases, 117–18; compression of architectural styles, 122–23; culture of, 101–4, 108; pithouses in, 110–13; thermodynamic trends and, 38; transition to Basketmaker III era, 113–15

Basketmaker III era: architectural complexity in North, 123–24; complexity increases, 117–18; compression of architectural styles, 122–23; pithouses, 121; pithouse villages, 146; traditional dating of architectural periods, 121; transition to, 113–15

baskets, 4, 38, 60, 77, 84–85, 101, 146, 151, 153, 183, 196, 203, 542

Beagle Channel, 20

beans, 2, 62, 65, 72, 88, 96, 102, 182, 184–85, 187, 198, 215, 229n1; enhancing nitrogen, 65; gardening expansion, 105–8; identifying, 76; as part of gardening progress, 82; planting, 74–75; preserving pocket garden, 108–9; and self-sufficiency, 164; steps in long path to sustainable gardens, 85; in women's horticulture, 78–79

beargrass (*nolina microcarpa*), 53

bear markets, manipulating, 178

beeweed, 120

"belonging," powerful psychological sense of, 186–87

big-game hunters, 23, 42, 226n4

big-game processing, calories consumed in, 76

Biocultural Approach to Human

Burialsfrom Chaco Canyon, New Mexico, A (Akins), 114

bison antiquus (extinct species), 16; and changing landscape in Alithermal period, 30–32; hunting in Paleo-Indian Southwest, 21–24

Blackwater Draw, 25, 28–30

blue smut (*huitzilocatchli*), 133

bow and arrow: power and efficiency of, 104–5; role in gardening expansion, 105–8

Braiding Sweetgrass: Indigenous Wisdom, Scientific Knowledge, and the Teachings of Plants (Kimmerer), 108

Bridges, Lucas, 57

British Empire, 7

broad-spectrum foragers: annual hours of, 74–75; descendants of, 78; and difference of thousand years, 98; and increased food competition, 55; increasing number of, 54; and Jemez Cave, 64–65; and pioneering gardening women, 62–63; seasonal camps of, 75; settling down, 59; shift to gardening, 71; social dynamics of, 46; toolkits and, 56

Broster, John B., 26

buckwheat, 38

Bunzel, Ruth, 217

burials, DNA study of, 157–59

Burton, Esther, 81, 120, 203–4

Butterfly Woman, 212

cactus, 30–31, 38, 41, 56, 63, 203

California, 2, 20, 26, 30–31, 95, 158

calories, 8, 21; and arrival of new millennium, 94–95; consumption of, 76; corn calories, 86; dietary calories, 47, 73, 76, 89, 109, 184, 195; essential role of, 1; expending, 22; food, 6, 27, 35, 74, 76, 80, 86, 89, 94, 106, 111, 125, 151, 154, 183; grass calories, 78;

heat calories, 17, 27, 110, 112; maize, 185; metabolic, 112, 117, 169; one work hour, 95; piñón nuts offering, 6; reducing, 17; saved, 17, 27; solar, 27

campsites, 56, 71, 78, 227n6; studying, 47; winter campsites, 39, 54

Casamero Community in the Red Mesa Valley of Northwestern New Mexico, The (Mathien), 189–90

Casamero (Great House), 189–90

catastrophe (in late 700s AD), 125

Chacoan Halo, 161

Chacoan World. *See* Chaco Canyon

Chaco Canyon, 1–3, 84; after Ice Age, 15–34; changes in Four Corners region, 35–51; complexity theory and, 147–48; designations, 161; emergence Chaco Canyon society, 131–49; farming labor patterns in, 91–97; genius and innovation of pocket gardens, 69–89; Great House era (875-1175 AD), 157–71; hammer of fate falling on, 201–4; Middle Archaic Period societies, 53–66; modifying landscapes as avenues of power, 177–88; origins of society in, 2; Pueblo women, 217–22; rhythms of Great House power, 189–200; semiarid empire dynamics, 149–56; shifting evolutionary balance, 99–115; small house vs. Great House society, 168–70; thermodynamic cultural state of, 205–6; ties to Mesa Verdeans, 165; tiptoeing on edge of chaos, 201–15; transition to Pueblo periods, 117–26; trip through, 128; wash district of, 166–68. *See also* Chaco Canyon society, emergence of; Chaco River; corn; food; gardening; pocket gardens; males; women gardeners

Chaco Canyon society, emergence of: Chaco Phenomenon, 148–49; complexity theory, 147–48; coprolite analysis, 133; early Great Houses, 136; formation of Chaco Canyon society, 145–48; heart of Chaco, 134; importance of Chaco River, 136–45; small house Chacoan architecture, 135; societal complexities in 800s AD, 131–35. *See also* Chaco River; corn; food; gardening; pocket gardens; males; women gardeners

Chaco Core: agriculture in, 162–63; designation, 161; wash district of, 166–68

Chaco Phenomenon, 148–49

Chaco Revisited (anthology), 157

Chaco River, 134–35, 161, 190; aerial views, 138–40; defining, 136; excavation near, 141–43; and Great House health benefits, 158; mesas of, 143–44; narrow ecotone carved out by, 140–41; physical environs, 144–45; as primary source of water, 143; and Pueblo Bonito, 137–40; small house vs. Great House society, 168–72; surveying, 131, 146; and still-undocumented expanse of wash district, 166–68. *See also* corn; food; gardening; pocket gardens; males; women gardeners

chamisa (rabbit brush), 77

chaos, tiptoeing on edge of: Butterfly Woman, 212; dawn of new world, 212–15; Edge of Chaos, 204; fate of small house farmers, 210; hammer of fate falling on Chaco Canyon, 201–4; post mortems on Great House culture, 206–10; reproductive dynamics, 210–12; thermodynamic cultural state, 205–6

chapalote, 71, 81, 85
chenopodia, 2, 33, 39, 47, 53, 57, 75–76
Chestnut, Elizabeth Akiya, 193, 217
Chetro Ketl, 134, 139, 144–45, 166
children, 92–93, 109, 146, 170, 202, 211–12, 220, 226n6; cooking duties of, 110; diet of, 43, 80; ethnographic trends, 83–84; extending information system, 61; forager dynamics, 46; hard work of, 45, 53; health of, 32, 43, 154–56, 170–71; high mortality of, 9, 87, 114, 197, 210–12, 230n6; hunter dynamics, 43–45; and pioneering gardening women, 62–63; uncovering remains of, 15–17; workloads of, 53. *See also* early women gardeners; women gardeners
Chimney Rock (Great House), 210
chokecherry, 38, 55
Christian calendar, dawn of, 87–88, 91, 100
Chuska Mountains, 69, 118, 144, 146. 160, 181, 194, 197
Chuska Valley, 69, 78, 154, 163
ciénagas (swamps or bogs), 73
climate, 2, 6, 217; of Altithermal period, 30–32; Clovis people and, 24–27; of Holocene Period, 21; influencing, 164; Pacific patterns of, 167–68; paleoclimate, 25; postglacial patterns of, 16–19, 28; predicting, 196; in San Juan Basin, 32–33; warming climate, 16, 27; wild shifts of mid-700s AD, 124
Clovis Era, 22–23
Clovis people, 19–20, 23–27. *See also* Altithermal Period; Early Archaic Period, changes during; Early Middle Archaic Period; Late Archaic Period; Middle Archaic Period; Pueblo periods, transition to; Terminal Archaic Period

"Coal Patch" towns (of Pennsylvania), 55
Coca Cola breakfast (meal), 103
Colt, Samuel, 104
Common Era, 3
competition, favoring, 48
complexity, increases in, 117–18
complexity theory, 137–38
cooking, efficiencies in, 109–10
coprolites, analyzing, 133
corn, 229n1; and Butterfly Woman, 212; consumption during Basketmaker II era, 103–4; cornfield reshaping daily life, 183–86; crops of, 92, 119, 177, 187, 193, 201; dietary benefits of, 86–87; diet of, 126, 155–56, 187–88, 201, 210; dried corn, 91, 96, 100–101, 110, 121, 151, 177, 187, 190, 196, 202, 205, 209; and dynamics of new power phase, 153–54; eating corn pollen, 159; and economic efficiency, 151; and engineered efficiency, 150–51; and expansion of gardening, 105–8; flint corn, 62, 76; fungus, 104; and gardening expansion, 105–8; of Great Houses, 193–94; identifying, 227n1; and "large field" planting strategy, 152; nutrition of, 104; pellagra and, 50, 133, 170–71, 198, 205–6, 211–12; and reproductive dynamics, 210–12; risk vs. reward in economy of, 187–88; sacred vs. secular, 193; seed corn, 73, 81, 83, 86, 100, 121, 124, 146, 149, 151, 159, 177, 187, 194, 219; small-cobbed corn, 2, 64, 71, 74–75, 96, 103, 157, 211; stored corn, 170–71; successful experimentation with, 100–101; train of farming, 152–53; unhusked corn, 150; varieties of, 62, 65, 82, 85, 88, 113, 169; and water-shaped "sharing networks," 118–19; and women's

health, 154–55. *See also* food; gardening; women gardeners
cotton seeds, 203
cottontail rabbits, 46, 61, 166
COVID-19, 84
critter fritters (meals), 41, 102
crops. *See* corn
culture: of Basketmaker II era, 101–3; post mortems, 206–10; rapid changes in, 65–66. *See also* gardening; Great Houses

daily life, cornfields reshaping, 183–86
Darwin-Lotka Energy Law, 1
Dawn Girl, infant, 15–16
designations, Chaco Canyon, 161
DeVore, Irven, 93
diet, foragers and, 70
dogs, ancient lineage of, 44
"Downtown Chaco," 161, 168–69, 181, 205
dryland garden districts, agricultural expansion in, 106
DuBois, Christine, 88
Dunmire, William W., 5
dynamics, 8

Early Archaic Period, changes during: critter fritters, 41; food efficiencies, 46–49; foraging cycles, 36–38; improving nutrition, 42; midsummer as challenging food season, 39–40; processing acorns, 50; producing "Chaco Phenomenon," 35–36; role of turkey, 49–50; seed collections, 38–39; social dynamics, 43–46; thermodynamics trends, 38; winter camps, 40–41; work roles of hunters vs. foragers, 42
early Basketmaker era. *See* Basketmaker II era
Early Middle Archaic Period, 5–6
early women gardeners: cultural evolutionary toward horticulture, 75–76; development of lifeway of, 74–75; ethnographic trends, 83–84; foreshadowing future, 82–83; overview, 69–70; power phase of 500 BC–200 AD, 85–89; power phases of, 79–80; progress in Basketmaker society (500 BC–500 AD), 81–82; rules, risks, and complexities of, 73–74; shift to Basketmaker horticulture, 80–81; shift to gardening, 71–72; sophisticated horticulture in Late Archaic Period, 78–79; steps in long path to sustainable gardens, 84–85; transition to Terminal Archaic Period (ca 500 BC–450 AD), 77–78. *See also* gardening; women gardeners
Eastern Asia, early peoples from, 15
ecology, post mortem on, 206
Edge of Chaos, operating on, 204
efficiency: cooking and, 109–10; cycles of, 9–10; dynamics and, 8; economy and, 151; engineering, 150–51; of pottery, 109–10; ratio of power to, 6–8; understanding concept of, 4–6
elephants, yearly amount of hunted, 29–30
environmental work. *See* postglacial warming, effects of
ethnography, trends in, 83–84
evolutionary balance, shift in scale of: anatomy of two innovations, 103–4; bow and arrow as power and efficiency, 104–5; gardening efficiencies, 103; gardening expansion, 105–8; labeling period as Basketmaker II, 101–3; life changes in San Juan Basin, 100; overview, 99; pithouses, 110–13; pottery and cooking efficiencies, 109–10;

evolutionary balance (*continued*)
    preserving pocket garden, 108–9; sharing networks as living cells, 107; storage practices, 101; successful corn experimentation, 100–101
exclusion, favoring, 48
expansion: of corn growing, 152; of gardening, 105–8; of Great Houses, 179–81; roads to, 160–62
Explaining Long-Term Trends in Health and Longevity (Fogel), 9–10, 198–99

fall hunt, 44–45
farm labor, patterns of: annual labor in San Juan Basi, 92–94; cementing "real estate" asset, 92; intensity, 91; monsoon seasons, 91–92; overview, 91–94; regional society advances, 94–95; thousand-year difference, 95–96; visions of new millennium, 96–97
female-created landscapes, genius of, 54. *See also* pocket gardens; women gardeners
female DNA, 226n7
females. *See* gardening; women gardeners
fertilizer, creating, 108. *See also* gardening
"figure five"–shaped throwing sticks, 18
Fogel, Robert W., 9–10, 198
food: competition for security of, 56–58; consequences of increased competition for, 55; efficiencies, 46–49; processing acorns, 50; risk of shortage of, 170; role of turkey, 49–50; sources of, 41. *See also* corn; gardening; women gardeners. *See also* corn; gardening
foragers: challenges faced in Late Archaic Period, 56–58; cultural trajectory in Middle Archaic Period, 51; decreasing plant diversity, 57; diet in, 70; era of technology and experimentation, 60–61; food efficiencies of, 48–49; and genius of female-created landscapes, 54; hunting-foraging, 7–8, 18, 24, 27, 50, 56, 71, 73, 80; hunting in fall, 24; intermarriage of, 62–63; and Jemez Cave, 64–65; lifeway development, 74–75; lowland foragers, 70; plant foragers, 23–24, 42, 80; pregnancy in, 70; rapid cultural changes of, 65–66; robust plant and foraging society of Middle Archaic Period, 53; role of turkey to, 49–50; social dynamics of, 46; studying campsites of, 47; tensions between hunters and, 54–56; upland foragers, 24, 70, 94, 160; work roles of, 42. *See also* broad-spectrum foragers
Ford, Richard, 47, 64
Fort Wingate, 153
fossilized remains. *See* coprolites, analyzing
Four Corners family, growth of, 85–89
Four Corners region, changes in: critter fritters, 41; food efficiencies, 46–49; foraging cycles, 36–38; improving nutrition, 42; procession acorns, 50; producing "Chaco Phenomenon," 35–36; role of turkey, 49–50; seed collections, 38–39; winter camps, 40–41; work roles of hunters vs. foragers, 42
Fragile Legacy of Well-Being, A (Stuart), 114
future, foreshadowing, 82–83

gardening: complexities of, 73–74; cultural evolutionary trajectory toward horticulture, 75–76; efficiencies in, 103; estimated rough dietary calories consumed in, 76;

ethnographic trends in, 83–84; expansion of, 105–8; fertilizer for, 108; foreshadowing future, 82–83; lifeway development, 74–75; long path to sustainable gardens, 84–85; opening San Juan Basin to, 59–60; power phase in period from 500 BC to 200 AD, 85–89; and power phases, 79–80; preserving pocket garden, 108–9; progress in Basketmaker society, 81–82; risks of, 73–74; rules of, 73–74; shift to, 71–72; shift to Basketmaker horticulture, 80–81; and sophisticated horticulture, 78–79; transition to Terminal Archaic Period, 77–78. *See also* women gardeners

gatherers. *See* broad-spectrum foragers; foragers; hunter-gatherers; hunters

Geib, Paul, 105

geography: Chaco Phenomenon, 148–49; complexity theory, 147–48; coprolite analysis, 133; early Great Houses, 136; formation of Chaco Canyon society, 145–48; heart of Chaco, 134; importance of Chaco River, 136–45; small house Chacoan architecture, 135; societal complexities in 800s AD, 131–35. *See also* Chaco River; corn; food; gardening; pocket gardens; males; women gardeners

Gibbon, Edward, 3

globemallow, 144, 166

gomphothere (elephant genus). *See* elephants, hunting

goosefoot, 38–40, 76

grama grass, 38

grandeur: of Great Houses, 179–81; at Pueblo Bonito, 181–83

grass seeds, 3–6, 30, 33, 45, 47, 53, 65, 76, 149, 156, 170

Greater Southwest, 2; after Ice Age, 15–34; changes in Four Corners region, 35–51; farming labor patterns in, 91–97; genius and innovation of pocket gardens, 69–89; Great House era (875-1175 AD), 157–71; in Middle Archaic Period, 53–66; modifying landscapes as avenues of power, 177–88; Pueblo women, 217–22; semiarid empire dynamics, 149–56; shifting evolutionary balance, 99–115; tiptoeing on edge of chaos, 201–15; transition to Pueblo periods, 117–26; understanding efficiency in relation to, 4–6

Great Houses, 1–3; Chaco Canyon designations, 161; Chaco Core agriculture, 162–63; competing subcultures, 163–64; construction episodes, 190–91; danger of insufficient protein, 170; defining society of, 168–70; and dynamics of semiarid empire, 149–56; eating corn pollen in, 159; ecology post mortem, 206–7; and Edge of Chaos, 204; emergence of, 136; enigma of Pueblo Pintado, 191–92; era of (875-1175 AD), 157–58; expansion and grandeur of, 179–81; expansions of, 160–62; health benefits of, 158; modifying landscapes as avenues of power, 177–88; ownership patterns between men and women, 192–93; post mortems on culture of, 206–10; power of, 193–94; power phase, 159–60; psychology post mortem, 208–9; Pueblo Bonito apogee, 193–96; in Red Mesa Valley district, 189–90; rhythms of power of, 189–200; robust and frail, 170; Shabik'eshchee village, 164–66; sharing corn with, 187–88;

Great Houses (*continued*)
  social complexities of elites of, 178–79; sociology post mortem, 207–8; status quo in, 196–200; thermodynamics post mortem, 209–10; tiptoeing on edge of chaos, 201–15; water technology of, 150, 166–69, 179, 183, 189, 191–92, 207, 218, 225n9. *See also* Pueblo Pintado (Great House)
Great North Road to the Underworld, 203

Haaland, Deb, 218, 222
health (of women), 154–56
heishe, 157, 230n1
herraduras ("horseshoes"), 179
History of the Decline and Fall of the Roman Empire (Gibbon), 3
Holland, John, 207
home camps, 57
homes. *See* Unit Pueblos
Hopi people, 169, 184, 197, 213–14, 216–19, 222, 227n3, 230n9
horticulture, 225n9; cultural evolutionary trajectory toward, 75–76; shift to Basketmaker horticulture, 80–81; sophistication of, 78–79
"House of Our Ancestors, The" (Marden), 157
"How are we doing?," question, 10–11
Huckell, Bruce, 47
Huckell, Lisa, 47
Huddleston, Roy, 110
huitzilocatchli. *See* blue smut
humans. *See* Chaco Canyon; Chaco Canyon society, emergence of; women gardeners
hunter-gatherers, 23, 32, 43, 92–93, 227n4
hunters: changing patterns of women's land use, 63–64; cultural trajectory in Middle Archaic Period, 51; food efficiencies of, 46–49; and genius of female-created landscapes, 54; intermarriage of, 62–63; lifeway development, 74–75; in Paleo-Indian Southwest, 21–24; role of turkey to, 49–50; social dynamics of, 43–45; tensions between foragers and, 54–56; work roles of, 42. *See also* gatherers

Ice Age, human culture after: Altithermal period, 30–32; climate charts, 24–27; Clovis people and climates, 24–27; cultural trends, 32–33; elephant hunting, 29–30; heading South, 20–21; overview, 15–19; Paleo-Indian Southwest, 21–24; second natural power phase, 27–29; territorial tensions, 33–34
Indian ricegrass, 2, 33, 36–38, 40, 54, 69, 77, 94, 97, 166, 170
innovations, anatomy of, 103–4
insufficient protein, danger of, 170
iron, deficiency in, 156

jackrabbits, 41, 46
Jemez Caldera, 118
Jemez Cave, 64–65, 71, 75, 94
jerked meat, 24, 36, 42, 48

Kantner, John, 189
Kaplan, Hillard, 63
Kauffman, Stuart, 107
Kearns, Timothy, 37–38, 74, 88, 95
Kelley, Klara B., 205
Kemper, Stella, 110
Kimmerer, Robin Wall, 108
Kin Bis sa'ani, 196
Kohler, Timothy, 205

labor, patterns of: annual labor in San Juan Basi, 92–94; cementing "real estate" asset, 92; intensity, 91; monsoon seasons, 91–92;

overview, 91–94; regional society advances, 94–95; thousand-year difference, 95–96; visions of new millennium, 96–97
Laguna Pueblo, 59
Lake Estancia, 28
Lamar, Cynthia Chavez, 218, 222
lance heads, 18–19, 25, 33
landscapes: of Altithermal Period, 30–32; architecture, 144; cornfields reshaping daily life, 183–86; genius of female-created landscapes, 54; Great House expansion and grandeur, 179–81; modifying, 177–78; powerful sense of "belonging," 186–87; Pueblo Bonito grandeur, 181–83; risk vs. reward in corn economy, 187–88; social complexities of Great House elites, 178–79
land use, changing patterns of, 63–64
La Plata River, 150
large animals, hunting, 21–24
Late Archaic Period, 59, 88; basis for storage baskets in, 4–5; challenges faced in, 56–58; changing patterns of women's land use in, 63–64; cultural evolutionary trajectory toward horticulture, 75–76; era of technology and experimentation, 60–61; and Jemez Cave, 64–65; lifeway development during, 74–75; pioneering gardening women in, 62–63; shift to gardening in, 71–72; sophisticated horticulture in, 78–79; thermodynamics trends in, 38
late Basketmaker era. *See* Basketmaker III era
Late Basketmaker Period, 84–85
late 700s AD, catastrophe in, 125
Lee, Richard, 93
Leffer, Lauren, 70
Lekson, Stephen H., 194, 206

lessons, learning, 125–26
light gardening, calories consumed in, 76
living cells, sharing networks as, 107
Loose, Richard, 191
Lotka, Alfred J., 1
lowland foragers, diet and pregnancy in, 70
Lukachukai Mountains, 60, 197
lysine, 50, 103–4, 159

maiz de ocho, 81–82
"Making Babies May Take 10 Times More Energy than We Thought" (Leffer), 70
males, 179, 206, 226n5; cornfield reshaping daily lives of, 183–86; dietary requirements, 156; DNA at conception, 226n7; and edible corn, 193–94; favoring sons over daughters, 45; food efficiencies of, 46–49; ownership patterns of, 192–93; powerful sense of "belonging," 186–87; and Pueblo women, 218–22; social dynamics of, 43–46; work hours of, 94. *See also* women gardeners
Man the Hunter (Lee and DeVore), 93–94
manos, 50, 131
Marden, Kerriann, 157
Marshall, Michael, 191
Martin, Debra, 154–55, 164
Martinez, Maria, 221
Mathien, Frances J., 189
mayorales, 84
Meat Eater. *See* Butterfly Woman
Mesa Verde (district), 3, 194, 197; architectural complexity in north, 123–24; catastrophe in late 700s AD, 125; competing subcultures, 163–65; hammer of fate falling on, 201–4; protecting storage niches in, 103

metabolic syndromes, 49–50
metate, 50, 81, 93
Middle Archaic Period: changing patterns of land use, 63–64; challenges faced in Late Archaic Period, 56–58; cultural trajectory in, 51; era of technology and experimentation, 60–61; female-created landscapes in, 54; Jemez Cave in, 64–65; opening San Jaun Basin to gardening, 59–60; pioneering gardening women in, 62–63; rapid cultural changes in, 65–66; regional societies in, 54–56; robust plant and foraging society of, 53; seasonal camps in, 54; sharing networks during, 61–62; thermodynamics trends in, 38
mid-700s AD, wild climate shifts of, 124
Mills, Barabra, 190–91
Minnis, Paul, 196
modern times, Pueblos of, 216
monsoon season, farming during, 91–92
Montezuma Valley, 125, 205
Morenon, E. Pierre, 151
Morrow, Baker, 108
"Most Critical Characteristics of Chaco, The" (Minnis), 196
Mother Nature, 27, 30
Mount Taylor, 163, 189
mtDNA, 2, 15, 226n6

narrow leaf yucca (*yucca glauca*), 53
National Park Service, 139, 141–42, 182
nature, watching, 225n6
networks, sharing, 61–62
New Mexico. *See* San Juan Basin
new millennium: arrival of, 94–95; visions of, 96–97
new world, dawn of, 212–15
niacin-producing plants, decline of, 86
nolina microcarpa. *See* beargrass

North, architectural complexity in, 123–24
North American Southwest, 1
Novotny, Judith, 191
nutrition, improving, 42

ojos (springs or seeps), 73
ollas (fired clay storage jars), 120
Ona-on-Ona raids, 57–58
ordinary families, cornfields reshaping daily life among, 183–86
ownership patterns (of women), 192–93. *See also* Great Houses

pack rats, analyzing, 42
Paisley Cave complex (Oregon), 18–19
paleoclimate, 24–27
Paleo-Indians, 28–29; and Altithermal period, 30–32; Clovis people and climates, 24–27; cultural trends of, 32–33; heading South, 20–21; hunting strategies of, 21–24; overview, 17–19; and second natural power phase, 27–29; territorial tensions, 33–34
Peach Springs, 78, 118, 128, 146
pellagra, contracting, 50
pemmican, 48
Peñasco Blanco, 134, 136–37, 161, 182, 186
piñón nuts, 6, 38, 40, 48, 54, 85, 94, 96, 102–3, 124, 156, 158, 163, 165, 187, 203, 213, 215
pithouses, 226n3; calories needed for constructing, 111; entryways, 112; floors of, 112; heavy poles for, 110; labor investment, 112–13; shaping of walls and roof, 111–12
plant foraging, calories consumed in, 76. *See also* calories
plant processing, 32, 51
playas, 16
Pleistocene, 15, 30
pocket gardens: cultural evolutionary

trajectory toward horticulture, 75–76; and dawn of new world, 212–15; ethnographic trends, 83–84; foreshadowing future, 82; gardening progress in Basketmaker society, 81–82; genius and innovation of, 69–70; lifeway development, 74–75; long path to sustainable gardens, 84–85; power phase in period from 500 BC to 200 AD, 85–89; power phases, 79–80; preserving, 108–9; rules, risks, and complexities, 73–74; shift to Basketmaker horticulture, 80–81; shift to gardening, 71–72; sophisticated horticulture, 78–79; transition to Terminal Archaic Period, 77–78; and wild climate shifts of mid-700s AD, 124
pollinators, 83, 108
Popular Science, 70
population, growth of, 89, 100, 113, 186
postglacial warming, effects of, 16–18
post mortems (on Great House culture): ecology, 206–7; psychology, 208–9; sociology, 207–8; thermodynamics, 209–10
potatoes, 6, 38, 228n4
pottery, efficiencies in, 109–10
power: cycles of, 9–10; ratio of efficiency to, 6–8; second natural phase of, 27–29
power phases: Basketmaker II to Basketmaker III, 113–15; dynamics, 153–54; and gardening, 79–80; Great House culture, 159–60; "large field" planting strategy, 152; in period from 500 BC to 200 AD, 85–89
precipitation: annual, 37, 83, 119, 184; predictable, 19, 148, 190; reliance on, 72; unpredictable, 72, 75, 97, 99
pregnancy, shortening spacing of, 210–12

pregnancy, foragers and, 70
primacy of land use, rule of, 73–74
procreation, dynamics of, 210–12
Prudden Unit Pueblos, 119–21
psychology, post mortem on, 208–9
Pueblo Alto, 134, 159
Pueblo Bonito (site), 92, 134, 144–45, 154, 166, 206, 211, 220; aerial views, 139–40; apogee of, 194–96; building stages at, 181; and Chaco River, 141–42, 168; construction episodes, 190–91; eating corn pollen, 159; economic efficiency, 151; emergence of, 136; grandeur of, 181–83; and Great House era (875–1175 AD), 157–58; Great House health benefits, 158; Great House power phase, 159–60; hammer of fate falling on, 201–4; Mesa Verde habitation at, 165; plan view of, 180; Pueblo Pintado echoing, 191–92; and sense of "belonging," 186–87; sharing corn, 187–88; take on, 137
Pueblo del Arroyo, 134, 138
Pueblo II period. *See* Pueblo periods, transition to
Pueblo I period. *See* Pueblo periods, transition to
Pueblo Peoples on the Pajarito Plateau: Archaeology and Efficiency (Stuart), 80
Pueblo periods, transition to: architectural complexity, 123–24; catastrophe in late 700s AD, 125; complexity increases, 117–18; compression of architectural styles, 122–23; lessons learned, 125–26; traditional dating of architectural periods, 121; Unit Pueblos, 119–21; water-shaped "sharing networks," 118–19; weather-shaped architecture, 118–19; wild climate shifts of mid-700s AD, 124

Pueblo Pintado (Great House), 128, 134, 161–62, 177; apogee of, 194–96; enigma of, 191–92
Pueblo women, ongoing knowledge of, 217–22
Puerco River, 153

rabbit brush, 77
rabbit hunting, calories consumed in, 76
"Raiding for Women in the Pre-Hispanic Northern Pueblo Southwest?" (Kohler and Turner), 205
real history, 8–9
Red Mesa Valley, 123, 150, 154, 162–63; Great Houses in district of, 189–90
Reed, Paul, 105
Reed, Paul F., 123
regional societies, tensions in, 54–56
regional society (800s AD), complexities of, 131–35
Reinhard, Karl, 133
reliance on precipitation. *See* precipitation
reproductive dynamics, upending, 210–12
rhus trilobata. See sumac
rhythms of Great House power. *See* Great Houses
Roosevelt, Eleanor, 178
Roosevelt, Franklin Delano, 178

sand dropseed, 38, 196
San Juan Basin, 167; and ancient lineage of dogs, 444; manual labor in, 92–94; architectural complexity in, 123–24; bow and arrow coming to, 104–5; changing patterns of women's land use, 63–64; climate in, 32–33; compression of architectural styles in, 122–23; construction episodes in, 190–91; cultural trajectory in Middle Archaic Period, 51; dawn of new world in, 212–15; dynamics of semiarid empire in, 149–56; and era of technology and experimentation, 60–61; female-created landscapes in, 54; female daily dietary requirements in, 156; fleeing, 205–6; foreshadowing future in, 82–83; gardening expansion in, 105–8; gardening progress in Basketmaker society, 81–82; innovations in, 103–4; and Jemez Cave, 64–65; opening to gardening, 59–60; overview of life changes in, 100; ownership patterns between men and women in, 192–93; pottery and cooking efficiencies, 109–10; power phase in period from 500 BC to 200 AD, 85–89; preserving pocket garden, 108–9; rapid cultural changes in, 65–66; regional society tensions, 54–56; rising population in, 33–34; rules, risks, and complexities of gardening, 73–74; sharing networks in, 61–62; shift to Basketmaker horticulture in, 80–81; shift to gardening in, 71–72; small house vs. Great House society, 168–70; status quo in, 196–200; studying self-extinction in, 57–58; thermodynamic trends in, 38; traditional dating of architectural periods in, 121; upending reproductive dynamics in, 210–12; upland and lowland foragers in, 70; weather-shaped architecture in, 118–19; wild climate shifts of mid-700s AD, 124; work hours rising in, 95–96. *See also* Chaco Canyon; corn; women gardeners
San Juan River, 150
Santa Fe Institute, 51, 96, 107, 137, 221,

230n7. *See also* living cells, sharing networks as
scion communities, 20
Sea of Cortez, 114, 118
seasonal camps, 32, 46, 54, 63, 75, 133, 143
seasonality, pronouncement of. *See* Altithermal (period)
self-extinction, studying, 57–58
self-help institutions, organizing, 61–62
self-storage, 48
Selk'nam Indians, 7
semiarid empire, dynamics of: corn diet, 155–56; dynamics of new power phase, 153–54; economic efficiency, 151; engineered efficiencies, 150–51; food train, 152–53; "large field" planting strategy, 152; overview, 149–50; women's health, 154–55
Seri peoples, 7
Seven Daughter of Eve, The (Sykes), 70
Shabik'eshchee, village, 164–66
sharing networks, 72, 99–100, 106, 146, 171, 187, 201, 207; and Chaco Phenomenon, 148–49; destruction of, 147–48, 187, 201, 207; disconnection of, 147–48; increased use of, 65; as living cells, 107; as organized self-help institutions, 61–62; in shape of water, 118–19
sheepsfoot, 38
Sherman, Jenny Lund, 184–85
Siberia, early peoples from, 15
single-family homes. *See* Unit Pueblos
Skunk Springs, 78, 128, 134, 146
small animals, hunting, 21–24
small house farmers, 167, 171, 186–87, 196, 199; fate of, 210
small houses, defining society of, 168–70
Smith, Mike, 110
social dynamics: of foragers, 46; of hunters, 43–45
sociology, post mortem on, 207–8
solar heat, rise of, 16–17

Sonora, 2, 7, 19–20, 23, 26, 29–31, 88
squash, 2, 78, 88, 92, 96, 121, 164, 182, 184, 198, 215, 229n1; in Basketmaker society, 82, 102; cultural evolutionary trajectory toward horticulture, 76; and development of gardening lifeways, 74; in expansion of gardening, 106–9; long path to sustainable gardens, 85; in robust plant and foraging society, 62–65; varieties of, 228n4
Stanford, Dennis J., 26
status: of men, 45, 85, 152; of women, 34, 43, 45, 85, 124
status quo (in Great Houses), 196–200
Stein, John, 191
storage practices, 101
Straits of Magellan, 20
subcultures, competition between, 163–64
suites. *See* Unit Pueblos
sumac (*rhus trilobata*), 53
sunflower seeds, 38, 57, 76, 102, 187
Sun Priest rooms, 181
sustainable gardens, steps in long path to, 84–85
Sykes, Bryan, 70

Talus Unit, 145
teosinte, 81
Terminal Archaic Period, 77–78
territory, tensions in, 33–34
thermodynamics, post mortem on, 209–10
thermodynamics, trends in, 38
tiered class systems, risks of, 83, 113–15
Tierney, Gail D., 5
Tierra del Fuego, 7, 16, 19–20, 57–60
tinajas, 20
toolkit, 56, 60
turkey, role of, 49–50
Turner, Kathryn Kramer, 205
two-room dwellings. *See* Unit Pueblos

Una Vida, 134, 136–37, 182, 186
United States, 8; cycles of power and efficiency in, 9–10; size of culture of, 10–11. *See also* Chaco Canyon; women gardeners
Unit Pueblos, 169; compression of architectural styles, 122–23; cornfields reshaping daily life, 183–86; emergence of, 119–21; learning lessons for, 125–26; Pueblo Bonito grandeur, 181–83
upland foragers, diet and pregnancy in, 70
upper piñon-juniper ecotone, 63
Upward Sun River, 15
Uttermost Part of the Earth (Bridges), 57

Varner, Bradley T., 110, 181
Vierra, Bradley J., 88
Vint, James, 181
Vivian, R. Gwinn, 92, 162

Waldrop, M. Mitchell, 204
wash district (Chaco), 166–68
water: carrying, 93; jars for, 103; technology of carrying, 150, 166–69, 179, 183, 189, 191–92, 207, 218, 225n9; water-shaped "sharing networks," 118–19. *See also* precipitation
Watson, Adam S., 92, 162
"we are struggling," elements of society, 114–15
weather-shaped architecture, 118–19
White Sands National Monument, 27
Wild Plants and Native Peoples of the Four Corners (Dunmire), 5

Wills, Wirt H., 166
Windes, Thomas, 190
winter hunt, 37
wolfberry, 38–40, 85, 106, 213
women. *See* children; early women gardeners; health (of women); ownership patterns (of women); status (of women); women gardeners

women gardeners, 1–3; after Ice Age, 15–34; changes in Four Corners region, 35–51; efficiency and, 4–6; farming labor patterns of, 91–97; genius and innovation, 69–89; Great House era (875-1175 AD), 157–71; in Middle Archaic Period, 53–66; modifying landscapes as avenues of power, 177–88; pioneering, 62–63; Pueblo women, 217–22; semiarid empire dynamics, 149–56; shifting evolutionary balance, 99–115; tiptoeing on edge of chaos, 201–15; transition to Pueblo periods, 117–26. *See also* corn; gardening; males; pocket gardens
women's health, corn and, 154–56

Yahgan Indians, 7
Y DNA, 2, 46, 226n4, 226n6
Youngr Dryas, 28
yucca glauca. See narrow leaf yucca

Zuni people, 197, 206, 212, 216–17, 219–20, 222, 227n10

About the Author

David E. Stuart's research and writing has long focused on history, culture, the natural environment, and how these factors influence the fabric and cycles of human societies. The dissonance between human behaviors and the forces and assets of natural environments motivate the constantly recurring phases of a culture's trajectory. We think of this as an economy, but in reality it is a never-ending battle between power and efficiency.

The author's writing was heavily influenced by the Stuart family environment in Philadelphia and the stunningly detailed remembered history of his lineage back to the ten hundreds AD in Brittany, as told to him by his great-aunts in a mix of English and Scots Gaelic. Memory has power in the life of human societies. So does wealth. Those Stuart aunts lived modestly but were rich in knowledge. Several years ago some of that knowledge was shared with Scottish scholars.

A second cherished family, the Densmore women, lived more prosperous lives in Philadelphia's Main Line. Their history was largely preserved in letters, formal documents, photographs, and in a book coauthored by his grandmother, Elsie Elizabeth Densmore, *John Hart Signer*. The author retains his status as a descendant of a signer of the Declaration of Independence. The Densmore women's brother, Benjamin Vandergrift Densmore, created advanced steamship engines, and later the world's first propjet engines used heavily in war. He was an inventor with stunning vision, but he died young.

The third family were the Dutch Vandergrifts, who took possession of Manhattan Island in the 1700s. Their sons and grandsons later dominated Mississippi River boat commerce until oil was found on their lands in

Pennsylvania. Their economic scion, J. J. Vandergrift, the fourth generation in America, created Standard Oil and is the author's third great-uncle. A trove of J. J.'s private documents are retained by the author.

The author's upbringing within these three families allowed him to see the fabric of America, its rich environment, and its stunning cultural ups and downs. From the trajectory of his families he derived the dynamics of power and efficiency, which have shaped his work.

www.ingramcontent.com/pod-product-compliance
Lightning Source LLC
Chambersburg PA
CBHW021347230426
43666CB00006B/438